P9-CAR-327

GALILEO'S GOUT

To Tatum,
My Scientist.
Love,
Leslie

Also by Gerald Weissmann

The Woods Hole Cantata
They All Laughed at Christopher Columbus
The Doctor with Two Heads
The Doctor Dilemma
Democracy and DNA
Darwin's Audubon
The Year of the Genome

GALILEO'S GOUT

Science in an Age of Endarkenment

GERALD WEISSMANN

BELLEVUE LITERARY PRESS
New York

First published in the United States in 2007 by
Bellevue Literary Press
New York

FOR INFORMATION ADDRESS:
Bellevue Literary Press
NYU School of Medicine
550 First Avenue
OBV 640
New York, NY 10016

This book was published with the generous support of Bellevue Literary Press's founding donor the Arnold Simon Family Trust, the Bernard & Irene Schwartz Foundation, and the Lucius N. Littauer Foundation.

PHOTO CREDITS

3, 15 (left), 19, 153: from the archives of the Accademia Nazionale dei Lincei, courtesy of the Secretariat of the Accademia, Rome, Italy; 163: Reproduced by kind permission of the Sir Francis Galton, F.R.S. website, http://galton.org/; 141: Courtesy of H.S. Lawrence; 55, 93, 157: Personal collection of the author.

Cataloging-in-Publication Data is available from the Library of Congress

Book design and type formatting by Bernard Schleifer
Manufactured in the United States of America
ISBN-10 1-934137-00-6 / ISBN-13 978-1-934137-00-0
FIRST EDITION
1 3 5 7 9 8 6 4 2

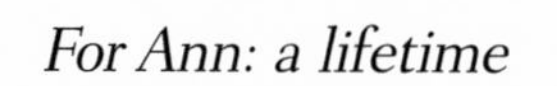

For Ann: a lifetime

Contents

We say pronounce, sentence, and declare that you Galileo . . . have rendered yourself in the judgment of this Holy Office vehemently suspect of heresy, namely . . . that the Sun is the center of the world and does not move from east to west and that the Earth moves and is not the center of the world. . . . Consequently we order that the book *Dialogue of Galileo Galilei* be prohibited by public edict. We condemn you to formal imprisonment in this Holy Office.

—The Holy Office of the Inquisition in Rome:
in re Galileo Galilei, Rome, August 22, 1633

Mr. Scopes, the jury has found you guilty under this indictment, charging you with having taught in the schools of Rhea county . . . a theory that denies the story of the divine creation of man, and teaches instead thereof that man has descended from a lower order of animals.

—Judge John T. Raulston: *Tennessee v John Scopes*,
Dayton, July 17th, 1925

Introduction: The Endarkenment

WE WHO ARE IN DEBT to rational science are living in the best of times and the worst of times, and that's not *A Tale of Two Cities*, but of two cultures. The prospects for each day in science have never been more splendid while our larger culture seems caught in a slough of unreason. In the last half century scientists have landed us on the moon, sampled Mars, and deciphered the human genome. Our new technology permits us to clone genes on chips and dial China from the Palm in our hand. The biological revolution has cracked new diseases as they arose (Lyme, HIV, SARS) and blunted the hurt of the old (cancer, cardiovascular disease, arthritis, and gout). We've doubled the longevity of fruit flies and roundworms in the lab and increased human life span in the developed world by a decade and a half.

Meanwhile, much of society at large is busy revisiting the days of the Inquisition v Galileo. The wars of religion are back, superstition threatens our schools, and Bible-thumpers preach that Darwin got it wrong. Our modern heritage of reason, formed in the Enlightenment, is becoming eclipsed by what a cynic might call the Endarkenment. It's no trivial matter when the editor of *Science*, Donald Kennedy, asks us whether it's "Twilight for the Enlightenment?"[1] Last year, for example, Gallup reported that 45 percent of Americans believe that God created human beings like us about 10,000 years ago. Indeed, less than a third of Americans believe that Darwin's theory is supported by scientific evidence, and just as many believe that evolution is just one of several, equally valid theories. A third of Americans believe that the Judeo-Christian Bible is the word of God to be taken literally, word for word.[2] One out of five Americans believe that the sun revolves about the earth!

For over a century in the lands of the West, the forces of faith and fact have largely observed an uneasy truce. Natural scientists were expected to steer clear of moral or religious matters, the clergy (save in the American South) were not expected to contradict the findings of Galileo or Darwin. But in our dangerous decade, the battle lines have formed again and this time they've been drawn around the globe. The trend—from Mississippi to Kabul to Jakarta and Jerusalem—is a return to revealed orthodoxy. Galileo may have been pardoned by his church, common descent accepted by a saintly pope, but the ancient tide of animist belief is rising again. We could have seen it coming. Fresh from the heady insights of the biological revolution, Jacques Monod warned us in his magisterial *Chance and Necessity* (1971) that

> Modern societies accepted the treasures and the power offered them by science. But they have not accepted—they have scarcely even heard—its profounder message: . . . a complete break with the animist tradition, the definitive abandonment of the "old covenant," the necessity of forging a new one. Armed with all the powers, enjoying all the riches they owe to science, our societies are still trying to live by and to teach systems of values already blasted at the root by science itself.[3]

Nowadays, in thrall to constituencies of unreason, zealots of all stripes are chipping away at evolutionary science. In our own country, "creationism" and "intelligent design" are now considered suitable topics for instruction in science, as if these notions were as testable as the perfect gas laws of Boyle ($pV = nRT$) or the Hardy-Weinberg equation ($p2 + 2pq + q2 = 1$) of population genetics. In the new spirit of Endarkenment, we ought not to be surprised that the Bishop of Vienna warned us against Jacques Monod:

> Scientific theories that try to explain away the appearance of design as the result of "chance and necessity" are not scientific at all, but, as John Paul put it, an abdication of human intelligence.[4]

Since the "abdication of human intelligence" might be defined as stupidity, it's clear to me that the truce has been shattered. Sad to say, it's possible that modern science may be at the stage of the arts in Quattrocento Florence, when poets and painters broke the mold of monkish severity in outbursts of wanton beauty. Savonarola responded by first scorning their intelligence and then burning their work. Schonhorn and the censors are gathering in the piazza.

So, for prelates and presidents, let's spell out the *facts* of evolution ever since Darwin: the facts of common descent and natural selection. Most scientists agree that evolution is no more a "theory" (in the popular sense) than is gravity. Evolution is based on a collection of facts that—like gravity—challenge Biblical notions of the nature of the universe and our very selves. The National Academy report on "Evolution and Creationism" reminds us that "In science, theories do not turn into facts through the accumulation of evidence. Rather, theories are the end points of science."[5]

The "theory" of evolution is based on six sets of facts that contradict any number of sectarian scriptures, and those facts alone should dictate what is taught as science in our schools:

1. The earth is about 4 billion years old.[6]
2. 5.7 million years ago we descended from an ancestor we share with chimpanzees.[7]
3. Homo erectus (our immediate ancestor) is anywhere from 1.8 to 0.3 million years old and people more or less like us (homo sapiens) arose in Africa 100,000 years ago.[8]
4. The fossil evidence for these observations has been validated by newer techniques of molecular biology, capped by the human genome project.[5]
5. Natural selection of random mutants (in portions of the genome that may or not be "hot spots") accounts for the emergence of new strains of viruses (influenza, HIV), microbes, and tumors. Natural selection and survival of the fittest have been demonstrated for the resistance of bacteria to antibiotics, of plants to herbicides, and of tumor cells to chemotherapy.[5]
6. Natural selection is the basis of immunity.[9]
7. Natural selection and survival of the fittest have been observed in humans as well. Ever since Rochelle Hirschhorn first described reversion to normal (both in DNA and clinical health) of a child with inherited adenosine deaminase deficiency,[10] reversion to normal, called "somatic mosaicism," has been shown in several disorders in which the revertant cells have a selective advantage.[11]

These facts of life science, directed by the physics and chemistry of DNA, turn out to obey laws as universal as those of Boyle's perfect gas. Just as $PV = nRT$ describes the behavior of gases in a rocket to the moon or an RPG in Fallujah; the ratios of G:C and A:T are equal in the DNA of fly and earthworm, mouse and microbe, prelate and president. The clock that sets the time to copy DNA was figured out in quahogs; the truth of molecular evolution is that we are such stuff as clams are made of.[12]

It would be reassuring for many of us were the lessons of Darwinian evolution simply a collection of tall stories we could take or leave at will—a tale of comfort or terror, of promise or warning, but tales after all of the mind, texts without bite. Marianne Moore described the world of poetry as composed of imaginary gardens with real toads in them. Well, I'm afraid that the facts of evolution are those of *real* gardens with real toads in them. They are not the baubles of one race, one gender, one class, or one Reich. They have been worked out by the buzzing of eager minds over complaints of the pious, the zealous, and the herbally inspired. Of course, evolutionary theory may be only one of several explanations for life on our planet, but it's the only theory that has held up against disproof. And however much we think we know of evolution today, it must be a minute fraction of what remains to be discovered tomorrow. Finally, I'd argue that the facts of evolution impose a kind of necessity on the chance of our imagination, they cut short many a tall tale. Experimental science is our defense—perhaps our best defense—against humbug and the Endarkenment.

Intelligent Design: Galileo and the Lynxes

> We say pronounce, sentence, and declare that you Galileo . . . have rendered yourself in the judgment of this Holy Office vehemently suspect of heresy, namely . . . that the Sun is the center of the world and does not move from east to west and that the Earth moves and is not the center of the world. . . . Consequently we order that the book *Dialogue of Galileo Galilei* be prohibited by public edict. We condemn you to formal imprisonment in this Holy Office.
>
> —*Judgment of the Holy Office of the Inquisition*, August 22. 1633[1]

> If once [the scientific] method were followed with diligence and attention, there is nothing that lyes within the power of human Wit (or which is far more effectual) of human Industry, which we might not compass; we might not only hope for Inventions to equalize those of Copernicus, Galileo, Gilbert, Harvey, and of others. . . .
>
> —ROBERT HOOKE, *Micrographia*[2]

> Senator Bill Frist of Tennessee, the Republican leader, aligned himself with President Bush on Friday when he said that the theory of intelligent design as well as evolution should be taught in public schools.
>
> —*New York Times*, August 20, 2005[3]

PRESIDENT BUSH AND DR. FRIST (Harvard Medical School, 1978) have been far more tolerant than their Inquisitional predecessors in dictating the texts of science. After all, they didn't put Darwin's *On The Origin of Species* on the Index of Forbidden Books or sentence James Watson to Gitmo. Nevertheless, most scientists remain persuaded that intelligent design is a euphemism for creation science: mutton served as lamb. One suspects that if the Holy See had not forgiven Galileo in 1992, our leaders might insist that American children be taught that the sun revolves around the earth. Taught alongside Galileo's heliocentric theory, of course.

That's not such a far-fetched notion. We might recall that it took twenty-three years for the Supreme Court (*Epperson v Arkansas*, 1968) to decide that it was unconstitutional for a state, "to prevent its teachers from discussing the theory of evolution."[4] We might also recall that in

1663 the Inquisition forced Galileo to recant. Riddled with gout and beaten in spirit, the seventy-year-old astronomer withdrew:

> My false opinion that the Sun is the center of the world and immovable . . . and that the Earth is not the center of the same and that it moves . . . and after it had been notified to me that said doctrine was contrary to Holy Writ.[5]

It wasn't the first time Galileo had been accused of heresy. Twenty years earlier Galileo had obtained evidence that Copernicus was right; he'd based his conclusions on telescopic images of sunspots projected on sheets of white paper. In the course of these heretical observations, he was drawn into the ranks of the world's first scientific academy, Prince Frederigo Cesi's Accademia dei Lincei (the Lynxes) founded in 1603. The Lynxes pledged themselves to "freedom of intellect, love of truth, and the true sources of human knowledge: not the dialectics of Aristotle, but reality based on reason, observation, and mathematics." As the sixth member of the Lynxes, Galileo sought Cesi's help in publishing his discovery, but ran into censorship problems with his *History and Demonstrations Concerning Sunspots and their Phenomena*. Cesi suggested publication of the sunspot observations in the form of letters addressed to another Linceian, Marcus Welser of Augsburg. Looking for an epigraph that might mollify the clerics, Galileo had counted on a passage on the Heavens from Matthew to do the trick: that didn't work. Cesi then advised Galileo to duck under the censorship bar by omitting scriptural references altogether, and a quotation from Horace was substituted.[6] With Horace at the mast, the sunspot letters were published in 1613 under the imprint of the Linceian Academy and soon "brought the question of the earth's motion to the attention of practically everyone in Italy who could read."[7]

It is not by chance that Galileo picked Horace's "Ode on Endurance" (Book III, ode II) to introduce his volume.

> *Virtus, recludens immeritis mori*
> *Caelum, negata temptat iter via,*
> *Coetusque vulgaris et udam*
> *Spernit humum fugiente pinna.*

> But Worth opens wide the gates of heaven
> to those who have done immortal things.
> Escaping the masses and damp earth itself,
> Worth raises the worthy on beating wings.[8]

ISTORIA
E DIMOSTRAZIONI
INTORNO ALLE MACCHIE SOLARI
E LORO ACCIDENTI
COMPRESE IN TRE LETTERE SCRITTE
ALL'ILLVSTRISSIMO SIGNOR
MARCO VELSERI LINCEO
DVVMVIRO D'AVGVSTA
CONSIGLIERO DI SVA MAESTA CESAREA
DAL SIGNOR
GALILEO GALILEI LINCEO
Nobil Fiorentino, Filosofo, e Matematico Primario del Sereniss.
D. COSIMO II. GRAN DVCA DI TOSCANA.
Si aggiungono nel fine le Lettere, e Disquisizioni del finto Apelle.

IN ROMA, Appresso Giacomo Mascardi. MDCXIII.
CON LICENZA DE' SVPERIORI.

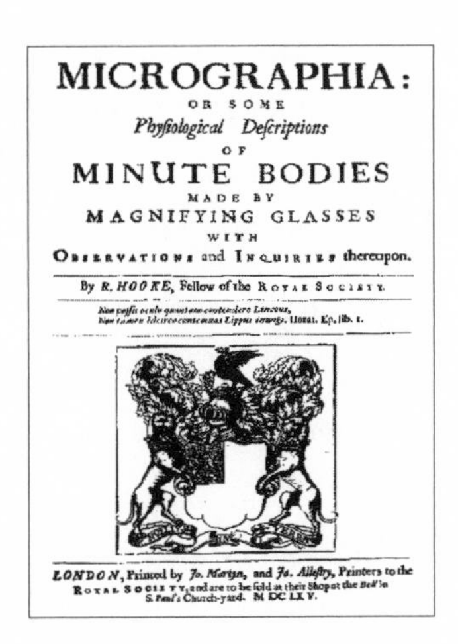

MICROGRAPHIA:
OR SOME
Physiological Descriptions
OF
MINUTE BODIES
MADE BY
MAGNIFYING GLASSES
WITH
OBSERVATIONS and INQUIRIES thereupon.

By R. HOOKE, Fellow of the ROYAL SOCIETY.

LONDON, Printed by Jo. Martyn, and Ja. Allestry, Printers to the ROYAL SOCIETY, and are to be sold at their Shop at the Bell in S. Paul's Church-yard. M DC LX V.

ABOVE LEFT: *Istoria: Title Page of Galileo's Monograph on Sun Spots (published by the Accademia dei Lincei, 1613).* ABOVE RIGHT: *Micrographia: Title page of Robert Hooke's work in which the term "cell" was first employed. (published by the Royal Society, London, in 1665.)*

Ever since the Renaissance, Horace (Quintus Horatius Flaccus, 65–8 BCE) has been a favorite poet of rational minds. One hopes he will remain so in the era of Dr. Frist, to remind us of what was lost when the Dark Ages eclipsed the glory that was Greece and the grandeur that was Rome. Happily, Horace was revived when monkish dogma yielded to the New Science of the seventeenth century. In his 1620 manifesto, *The Great Instauration*, Francis Bacon echoed Horace:

> For I admit nothing but on the faith of eyes, or at least of careful and severe examination, so that nothing is exaggerated for wonder's sake, but what I state is sound and without mixture of fables or vanity.[9]

The "faith of the eye" is a paraphrase of Horace's allusion to the eye of the lynx, the animal with the keenest vision.[6] The faith of the eye, a reliance on the evidence of things newly seen, connects Galileo Galilei with Robert Hooke, the microscope with the telescope, and Italian fossils with the origin of species. The faith of the eye also connects the first two societies of experimental science, the Lincei of Rome (1603) with the Royal Society of London (1662). How apt, then, that Horace also provided the motto of the Royal Society. *Nullius in Verba*, the motto of the Royal Society to this day, derives from Horace's First Epistle, *To Maecenas,* a skeptic's creed if ever there was one:

Ac ne forte roges quo me duce, quo lare tuter,
nullius addictus iurare in verba magistri . . .

And if by chance you ask what direction I follow
I am bound by the words of no master . . .[10]

Hooke's *Micrographia,* published by the Royal Society in 1665, sports on its title page a further quote from Horace's First Epistle:

Non possis oculo quantum contendere Linceus,
non tamen idcirco contemnas lippus inungui . . .

While your eyes can't match the vision of a lynx,
You'll salve them with medicine when swollen.[10]

The epigraph is a clear tribute by the Royals to their predecessors, the Lynxes of Rome and of their pledge. Indeed, Hooke's *Micrographia* is a record of observations made of creatures great and small with the microscope, an instrument that extends the range of the human eye to rival that of the lynx. When Hooke introduced the results of his research to the public, he again paid homage to Galileo, the Celestial Observer:

> For the members of this Society having before their eyes so many fatal instances of the errors and falsehoods, in which the greatest part of mankind has so long wandered . . . have begun anew to correct all *Hypotheses* by sense, as seamen do their *dead Reckoning* by Celestial Observations; and to this purpose it has been their principal indeavour to enlarge & strengthen the *Senses* by *Medicine*, and by such *outward Instruments* as are proper for their particular works. By this means they find some reason to suspect that those [phenomena] confessed to be occult, are performed by the small machines of Nature.[11]

The outward instrument Hooke used to examine the small machines of nature in *Micrographia* was the compound microscope. With it Hooke first saw that living matter (*i.e.*, cork) was partitioned into smaller units that resembled the cells of monks. That's how cell biology began. The instrument he used was a version of one that Galileo had introduced fifty years earlier and called an occhialino, and for which Galileo's fellow Lincean, Giovanni Faber, had proposed the name we recognize today, *"Microscopium"* (1625).[12] Later that year, Francesco Stelluti, another Linceian, made the first engravings of natural structures obtained by means of the new instrument: detailed and still ravishingly beautiful images of bees in his *Melissographia.*[13] Stelluti, more politic by far than

Cesi or Galileo, dedicated his folio to Pope Urban VII of the Barberini family, whose family *distintivo* (coat of arms) sported honeybees.

Hooke appreciated that the Lynxes were first on the micrographical map and gave "Francisco Stelluto" full credit, in the course of correcting Stelluti's notions of how fossils arose.[14] With Cesi, Stelluti had collected many hundreds of specimens of fossil woods, animals and pyrites, and on Cesi's death—the year of Galileo's trial—published a small volume based on their joint studies, *Trattato del Legno Fossile Minerale*. Stelluti claimed that fossils were not generated from any plant or living thing, but only from a type of clay, which slowly became transformed into wood.[15] Stelluti's notion that fossils were never alive may—or may not—have been designed to placate the Inquisition. We'll never know. Intelligent Design had laid down the law: life came from clay. In contrast, Hooke correctly deduced that living things fossilized in the Umbrian rocks—or elsewhere—might be traces of vanished life: life turned into clay or stone. We now appreciate that the Stelluti/Hooke controversy was the beginning of the search for the origin of species: what we now call the fossil evidence for evolution. That line of investigation led to Darwin in Down House and to John Scopes down home in Tennessee.

In the hands of Linceans and Royals, telescope and microscope were the tools that made it possible to decipher the small machines of nature. Nowadays, the tools we use are legion. Look at the abundance of figures and tables printed in any issue of a modern journal of science. What a feast is presented to "careful and severe examination" à la Hooke. Look at those gene sequences spilled from machines, the fluorescent transgenes traced in living cells, the X-ray diffraction patterns or those NMR squiggles, the microchips and DNA sequences. Our charge as scientists is to keep the faith of the eye as opposed to exaggeration for wonder's sake.

In the sphere of art and belief, a sense of intelligent design may have inspired masterpieces like the Bach B Minor Mass or the Sistine Chapel. But, ever since Galileo, the notion of intelligent design has lost its place in the sphere of science. *Nullius in verba magistri,* say we.

Galileo's Gout

ONE OUT OF FIVE AMERICANS MAY remain convinced that the sun revolves around the earth, but our millennial Endarkenment cannot eclipse the star of Galileo (Galileo Galilei, 1564–1642). His legacy is scattered about the universe. In November of 2002, one of NASA's longest-running missions came to an end when the Galileo spacecraft, launched in 1989, made its final orbit of Jupiter, the planet whose four moons Galileo first described in 1610. With its ice-capped poles, one of those moons, Europa, seems a better candidate than Mars as habitat for extraterrestrial life.[1] Earlier that year, Europe—the continent—filed its answer to the American Global Positioning system; it shot a satellite into orbit and called it Galileo.[2] Galileo's own stock rose when physicists ranked two of Galileo's experiments among "Science's 10 Most Beautiful Experiments."[3] Our decade also marked the 400th anniversary of the *Accademia dei Lincei* (Rome, 1603) the world's oldest scholarly society, of which Galileo was dare we say—the star. To mark the occasion, a delightful volume by Columbia's David Freedberg, *The Eye of the Lynx* showed how the new worldview of Galileo and his Linceians was an impetus for London's Royal Society (1662) and Colbert's *Académie des Sciences* (1666).[4] Finally, a definitive exhibition on Albert Einstein at the American Museum of Natural History credited Galileo with anticipating the notion of space/time as in *The Assayer* (1623):

> whenever I conceive of any material or corporeal substance, I immediately feel the need to think of it as bounded, and as having this or that shape; as being large or small in relation to other things, and in some specific place at any given time; as being in motion or at rest; as touching or not touching some other body; and as being one in number, or few or many.[5]

Engraving of Galileo's Portrait by Ottavio Leoni (1624)

Galileo has affected the broader culture of our new millennium as well. Philip Glass and Mary Zimmerman premiered their opera *Galileo* in Chicago, London, and New York. This downtown version of the seventeenth-century face-off between obstinate fact and adamant faith gave British critics the last word.[6] "Glass and Zimmermann not only insult the intelligence of their audience with their profoundly banal efforts, but also trivialize one of the greatest minds of the renaissance." The PBS version of this seventeenth-century World Series was billed by *Nova* as "Galileo's Battle for the Heavens."[7] Fact lost to faith in four games: the Cardinals won handily, presumably because of home-field advantage. As the *Boston Globe* delicately put it, "Galileo hobbles around his study, laments his illnesses, and ruminates on how he can open the eyes of the Church to the true nature of the universe."[8] In real life, Galileo had the last word in his famous letter to Castelli:

> Thus it appears that physical effects placed before our eyes by sensible experience or concluded by necessary demonstrations should not in any circumstances be called in doubt by passages in Scripture that verbally have a different appearance. Not everything in Scripture is linked to such severe obligations as is every physical effect.[9]

Galileo had previously tried to soft-pedal his support of the Copernican view of the heavens, but to no avail (see "Galileo and the Lynxes"). As he feared, his sunspot letters gained acute attention by the Inquisition, and the battle was on between Church and the State of Nature.

I suspect that current interest in the old match-up of Scripture v Physics has less to do with Galileo than with more recent contests such as Frist v Darwin or Scalia v Freud, Marx, and Einstein. Like most scientists I had considered the issue moot ever since Pope John Paul II "forgave" Galileo in 1992 and, if the Dover decision is a bellwether, there is hope that even the Bible-thumpers will eventually "forgive" Darwin. Be that as it may, I'm interested in the question of why Galileo was hobbling painfully about in his confinement.

Prompted by *The Eye of the Lynx*, and by Dava Sobel's deservedly popular book *Galileo's Daughter*,[10] I've concluded that Galileo was not only a victim of the Inquisition, but of saturnine (lead-induced) gout. His biographers, from Drake to Sobel agree (a) that Galileo was indeed hobbled by "gouty" arthritis most of his days; (b) that he suffered from frequent kidney stones, bloody urine, and renal infections, of which he eventually died; (c) that he suffered since midlife from abdominal pains, usually ascribed to a hernia, for which he wore a heavy iron truss; and finally (d) that not only was he a lifelong heavy drinker but presided over his own cottage winery, where metal-bound casks of wine often turned to vinegar. Physical proof? Although there are about half a dozen contemporary portraits of Galileo extant, only one shows his hands.[11] Painted late in Galileo's life by his neighbor from Passignano, Domenico Cresti, the left hand shows what any arthritis doctor would recognize as the signs of chronic gout: tophi, interosseal atrophy, and flexion contractures. (Chalky deposits on knuckles, loss of flesh between fingers, and clawlike hands.)

Nowadays we appreciate that gout is due to crystals of uric acid dropped out of solution from tissue fluids and deposited in joints, skin, and kidney. In most cases this is due to a heritable flaw, to dehydration, to weight loss, or plain old alcoholism. In saturnine gout, those crystals of uric acid precipitate because there is too much lead in the body. And although Saturnine gout has been around since classical times,[12] no one appreciated that gout might be due to lead-laced wine or aqueduct water until the eighteenth century. In alchemical notation, *lead* was the sign for Saturn, furthest and slowest of all the planets. Astrologers believed, and nowadays alternative medical folks *still* believe, that those born under the sign of Saturn are cold, sluggish, and of morose temperament. In the eighteenth century, "sluggish" and "morose" became linked to "overweight" to describe the saturnine, plethoric peers who were drunk as lords on lead-laced port.

Galileo was only one of many comfortable, wine-swilling sages of the sixteenth and seventeenth century who came down with saturnine gout, manifest as kidney stones, podagra (gout of the big toe), or both.[12] Among statesmen afflicted by gout were Philip II of Spain, Charles V of the Holy Roman Empire, and *all* the Medicis. Giovanni, father of the great Cosimo, was crippled by the disease, while Cosimo himself became crippled later in life. His son Piero (1464–1469), father of Lorenzo the Magnificent, was called "il Gottoso." Lorenzo's own son, Pope Leo X, a satyr if ever there was one, suffered from swollen joints and kidney stones. Gout struck the period's greatest artists (Michelangelo, Peter Paul Rubens, Claude Lorrain)[13] poets, and scientists alike. Gouty John Milton is said to have sipped local wine with Galileo on a visit to Acetri. Gout went on to kill William Harvey, who in 1602 had received his medical degree in Padua with Galileo's doctor. Podagra tormented Isaac Newton and felled Wilhelm Leibniz, who died of gout after a week of stomach "colick" (known as lead colic in later years). It's probably no accident that this first wave of gout among the rich and famous prompted publication in 1643 of Guillaume de Baillou's first distinction between "rheumatism" (rheumatic fever) and true gout.[14] Good wine for the classes, strep throats for the masses, or, as Lord Chesterfield quipped "Gout is the distemper of a gentleman, rheumatism is the distemper of a hackney coachman."[15]

We learned about lead-induced gout from the great epidemic of saturnine gout that followed wholesale export of fortified Mediterranean wines to England and its colonies in the eighteenth century.[16] In an effort to stem the tide, so to speak, of French wine imports, the English foreign minister, Methuen, signed a treaty with Portugal in 1703 that permitted fortified wine (port and madeira) to be imported from Portugal at a fraction of the cost of the French products. He had inadvertantly struck a blow not only to French exports, but also at English kidneys. Wines such as port, madeira or malaga are fortified by adding distilled alcohol or brandy. But the stills of the Portuguese in the eighteenth century ground exceedingly fine: they were lined, or joined, by lead. Lead leached from the stills found its way into the finished product and remained in the bottle until drunk. The alcohol was added to fortified wine not only to give it more kick, but to prevent spoilage and deterioration on storage. It worked: the port lasts, and so does the lead. Indeed, recent spectroscopic analyses of rare, eighteenth-century fortified wines—on sale at Christie's—showed them to contain 1,900 micrograms/L, as opposed to the 180 micrograms/L in their present-day counterparts.[16]

In any event, by 1825 the English were importing 40,277 tons of fortified wine each year. As among the seventeenth-century toffs, the gout epidemic in England took its greatest toll among the middle and upper classes, for the poor drank mainly gin and beer. The well off, who could afford it, drank fortified wine à la George IV; they were fit subjects for cartoonists like Rowlandson and Bunbury, who lampooned a ruby-nosed, red-footed gentry lapsing into tophi and tremors. Whatever the century, gout was only one consequence of chronic lead intoxication, or plumbism.[17] The others are kidney disease, abdominal pain, *i.e.*, "lead colic," constipation, headache, fatigue due to anemia, and early, severe dental decay (the "lead line" on teeth).[18–22]

How does lead get into wine of the sort that Galileo drank? While natural variations in the lead levels of ground water and irrigation systems contribute to the lead content of grapes, the major source is from the winery itself. Lead from solder (the cooper's work) leaches from the barrel, a process hastened when wine is spoiled or acidified; lead seeps from bottle foils through wet corks. And as with port, lead also leaches from soldered distillation tubes used to fortify grappa or brandy. Galileo was a sitting duck for lead: he drank from the rotting barrels of his own winery and fortified spirits at the tables of the Medici and the Lincei.[23, 24] Direct evidence for the origin of Galileo's gout comes from his daughter, Suor Maria Celeste (1600–1634). Thanks to Dava Sobel and Rice University's "Galileo Project," his daughter's letters to Galileo are available online.[24] For ten years, from 1623–1633 through triumph and trial, honors and recantation, the daughter commiserated with her father's gouty aches and pains: *"This morning I learned from our steward that you find yourself ill in Florence, Sire: and because it sounds to me like something outside your normal behavior to leave home when you are troubled by your pains, I am filled with apprehension, and fear that you are in much worse condition than usual."* [17 August 1623] It is only late in the game, as Galileo is detained by the Inquisition in Rome and Siena, that we learn how those pains arose: *"I implore you not to confuse yourself with drink, as I hear you have been doing."* [21 May 1633] *"I am sorry that your pains give you no respite, although it seems almost requisite for the pleasure you take in drinking those excellent wines to be counterbalanced by some pain, so that, if you refrain from imbibing large quantities, you may avoid some greater injury that could be incurred by drinking."* [4 June 1633]

During his detention, Maria Celeste manages his home and wine cellar in Arcetri, from which much of Maria Celeste's own wine supply is derived: *". . . meanwhile, we have recovered one barrel from the farmers here, and had it put into the cask which formerly held that spoiled wine; . . . at my behest, Signor Rondinelli had a word with the blacksmith about the 3 barrels that he owes us, and brought back his solemn promise on that score."* [8 October 1633]

I have a hunch that poor Maria Celeste was herself a victim of plumbism. Her letters spell out a litany of chronic headache, anguishing tooth pain, and recurrent bouts of intestinal "obstructions" or "blockages." In those years between 1623 and March 1634, when she succumbed to a bloody bout of explosive diarrhea, she was, *"so accustomed to poor health that I hardly think about it, seeing how it pleases the Lord to keep testing me always with some little pain or other."* [23 November 1623] At age 28 (!) she wrote her father *"I tell you that I am following the doctor's orders by not observing Lent, and that, being already mostly toothless at my age,"* [25 March 1628]. These troubles continued most of her short, devoted life.

But her main problems were abdominal, and recalcitrant to herbal treatments or a *". . . pleasant purgative, in order to try to remove an obstruction that has troubled me (aside from my usual ailments) for the past six months."* [21 January 1630] She was still blocked by May: *"I feel reasonably, but not entirely well, since I am still taking the purgative on account of my blockage."* [25 May 1630] She remained blocked in February of 1631 *"as far as my long-standing blockage, however, I believe that will require an effective cure at a better time."* [18 February 1631] More GI distress, more wine: *"I am returning two empty flasks, and truly, in this slump I have had, were it not for your white wine, Sire, things would have gone much worse, since I sustain myself on pap and soup, which have not hurt me for being made with such good wine."* [July undated 1631]

More of her father's wine was needed for her fellow nuns—who were themselves unusually prone to colic and stoppages: *"If you would be so kind, Sire, as to send a flask of well aged red wine for her, I would be most grateful, because our wine is very harsh, and I want to try, in any small way I can, to help her to the last."* [14 January 1630] The abdominal epidemic in her convent waxed and waned for several years: "*We are all feeling fine, except Suor Luisa, who for the past three days has been suffering on account of her stomach, although not as severely as at other times.*" [30 April 1633] Maria Celeste tells us that her fellow nuns got their wine—and perhaps their symptoms—from Galileo's Arcetri casks: *"As for the wine that was decanted . . . we have finally found willing takers for the 3 barrels worth that we must give away. Suor Arcangela will not have to be begged to help them along."* [3 August 1633]

Poor Maria Celeste succumbed to her disease a half year later. She didn't live long enough to suffer from gout: Hippocrates taught us that "women do not get the gout unless their menses be stopped." It's my feeling that they ought to make Maria Celeste the patron saint of wineries the world over. As for Galileo, he was no saint, but he's still up there in the sky.

Swift-boating Darwin: Alternative and Complementary Science

AMERICAN SCIENTISTS BREATHED a collective sigh of relief in December 2005 after Judge John Jones 3rd ruled against teaching intelligent design (ID) in the classrooms of science. "The overwhelming evidence is that Intelligent Design is a religious view, a mere re-labeling of creationism and not a scientific theory," Jones declared in his 139-page decision, issued in Dover, Pennsylvania. "It is an extension of the Fundamentalists' view that one must either accept the literal interpretation of Genesis or else believe in the godless system of evolution. . . . The evidence presented in this case demonstrates that [intelligent design] is not supported by any peer-reviewed research, data or publications."[1, 2] But Dover isn't over.

Proponents of ID stubbornly refused to give up their campaign: "A thousand opinions by a court that a particular scientific theory is invalid will not make that scientific theory invalid," claimed Richard Thompson, of the Thomas More Law Center, a group long devoted to Swift-boating Charles Darwin. The center had previously boasted that when ID had been inserted into the Dover science curriculum "Biology students in this small town received perhaps the most balanced science education regarding Darwin's theory of evolution than any other public school student in the nation."[3] Robert Crowther, director of the Discovery Institute, a Seattle based think tank, so to speak, complained to the *New York Times* that Judge Jones's decision "asserts the factually false claim that ID proponents haven't published peer reviewed papers. A number of peer-reviewed papers and books are listed on the Discovery Institute website at *www.discovery.org/csc/*."[4, 5] William Demski, a mathematician and

Fellow of the Discovery Institute, insisted that "I think the big lesson is, let's go to work and really develop this theory and not try to win this in the court of public opinion . . . the burden is on us to produce."[1] Demski, you've got a heck of a job to do.

The Web site of the Discovery Institute reveals that the "peer-reviewed evidence" for ID consists of *four* articles. Each presents a theoretical argument that fails the test of experimental validation. Each has appeared in a publication devoted to pure speculation, including the occasional *Proceedings of the Biological Society of Washington;* the thrice-yearly Italian/Indian review *Rivista di Biologia/Biology Forum;* the yearly *Dynamical Genetics;* or the every-other-year *Proceedings of the Second International Conference on Design & Nature.*[5] We can conclude that active investigators of ID do not stoop to frequent or rapid publication. Nor does prestige dictate their choice of venue. High Wire Press,[6] where 70 of the highest cited journals have been archived since 1948, lists Darwin's "natural selection" in 271 titles; almost all are experimental accounts. "Intelligent design" appears in the title of but a single effort, a conjectural review published in the *Journal of Theological Studies.*[7] Alas, ID loses out to another system of alternative science: "Mesmer" or "Mesmerism" appears in 20 titles: each is devoid of experimental promise.

Alternative and Complementary Science

Although ID clearly lacks support in the literature of experimental biology, intelligent design remains a powerful notion that is no longer limited to extreme Fundamentalists. ID may be coming soon to a science classroom in your neighborhood. At least ten states have legislation pending that would declare ID an alternative, or complementary, view to Darwinian evolution. And while Darwin's "theory" of evolution is as well accepted by scientists as the heliocentric theory of Galileo or the gravitational theories of Newton, it's easy to see why true believers resist the facts of common descent and natural selection. As Judge Jones decided, Darwinian evolution clearly contradicts "the literal interpretation of Genesis" and resolving that contradiction is difficult at best.

But, I'm afraid that not only creationists or evangelists have questioned the experimental basis of science. The notion that there are alternative or complementary systems of medicine other than those based on the laws of physics and chemistry has swept not only day-time television, but captured the hearts and minds of our legislatures, our elite universities and found a home on the campus of the NIH.[8,9] The National Center for Complementary and Medicine explains why it is funding work based on Ayurvedic notions of animal magnetism:

> This vital energy or life force is known under different names in different cultures, such as qi in traditional Chinese medicine, ki in the Japanese Kampo system, doshas in Ayurvedic medicine, and elsewhere as prana, etheric energy, fohat, orgone, odic force, mana, and homeopathic resonance.[9]

I'm afraid that our current tolerance of homeopathic, chiropractic, Ayurvedic, holistic, crystal-based or aroma-driven modes of healing has helped to clear the way for the alternative or complementary science of intelligent design. Once advocates of folk-based remedies persuaded the public that there are alternative or complementary explanations of what ails us, why not accept faith-based alternative or complementary explanations for how we came about? If the laws of chemistry and physics (eg. PV=nRT) need not apply to medicine, why should we rely on the laws of evolution such as that of natural selection or the Hardy Weinberg equation?

We live in an open, diverse society, disdainful since the 1960s of the hard facts of science. That disdain has both intellectual and religious origins: the intellectual roots are chiefly French, the religious roots American. On the one hand, the best and the brightest among us have been tutored in what Nicholas Kristof of the *New York Times* called the "Hubris of the Humanities."[10] We have breathed the air of a postmodern era in which melancholic disciples of Michel Foucault proclaim "the end of our great faith in Progress."[10] On the other hand, American science teachers in Evangelical schools teach students that God created the world in six 24-hour days.[11] No wonder that only 40 percent of Americans believe in evolution and that only 13 percent know what a molecule is.[9] There are more professional astrologers than astronomers (10,000 v 800) in our country, more who preach metaphysics than physics (422,000 ministers v 16,000 physicists).[12]

New York and Nancy

The Dover decision was a landmark for those who value fact over faith in the realm of science; happily, there have been other such moments. Two public exhibitions (that happened to coincide with the

Dover decision) reminded me of episodes as important to the life of science as that ruling by Judge Jones. The American Museum of Natural History mounted a comprehensive and compelling show on the life, work, and everyday impact of Charles Darwin. Illustrating Theodosius Dobzhanky's aphorism that nothing makes sense in biology except in the light of evolution,[13] it attracted crowds of every age and hue in New York, and traveled to acclaim in Chicago, Toronto, and London.

Simultaneously, France celebrated 250 years of Light and the Enlightenment in a splendid exhibition at Nancy that served to illustrate Denis Diderot's argument that one can't traffic in metaphysics or morality without understanding the facts of natural science.[14] It was organized by Jean-Pierre Changeux, the dazzling polymath of the Collège de France and our century's best friend of reductive science. The exhibit featured original texts, scientific artifacts, prints, and masterly paintings that documented the triumph of scientific light and reason over the forces of "*obscurantisme*" (the Endarkenment). Exhibits ranged from Newton's spectrum to Mme du Châtelet's equations to modern images of nerve conduction. It was good to see that in December the galleries at Nancy were as packed as the corridors in Manhattan or that courtroom in Dover.

Huxley and Wilberforce

The Darwin show in New York called attention to the famous "monkey" debate at the Oxford Museum of Natural History on June 30, 1860.[15] The great debate began with a two hour-long treatise by Professor John William Draper, of the Medical Department of New York University, invited as the major American champion of Darwinist thought (see "Reducing the Genome"). Thomas Huxley remembered Dr. Draper at Oxford as "of course bringing in a reference to the *Origin of Species* which set the ball rolling." The details of what followed are controversial, but one exchange is engraved in the story of evolution.

Bishop Wilberforce—a premature televangelist and equivocal success as a mathematician—spoke next and taunted Huxley by asking if it was on his grandmother's or his grandfather's side that he was descended from apes. Huxley replied, famously, "I asserted, and I repeat, that a man has no reason to be ashamed of having an ape for a grandfather. If there were an ancestor whom I should feel shame in recalling, it would rather be a man, a man of restless and versatile intellect, who, not content with an equivocal success in his own sphere of activity, plunges into scientific questions with which he had no real acquaintance, only to obscure them by an aimless rhetoric, and distract the attention of his hearers from the real point at issue by eloquent digressions and skilled appeals to religious prejudice."[15]

Franklin and Mesmer

Changeux's exhibit at Nancy displayed an earlier memento of a similar setback for Unreason. It was the report of a Royal Commission appointed by Louis XVI to look into the activities of Franz Anton Mesmer (1734–1815). Mesmer had intruded his notion of animal magnetism into the highest level of French society. His doctrines leaned heavily on the Swedenborgian notion that matter is a subset of Mind, a notion antithetical to the teachings of the *philosophes* and the Academy itself. As we've learned from Robert Darnton, there was a disturbing connection between the rise of Mesmeric belief and the end of the Enlightenment in France.[16] Mesmer taught that disease resulted from various obstacles to the flow of a magnetic "fluid" or vital energy in the body. In a Mesmeric session, patients sat about in circular tubs and communicated the fluid by means of a rope looped about them all and by linking hands to form a mesmeric "chain." Soft music, played on wind instruments, a pianoforte, or a glass harmonica reinforced the waves of ethereal energy that "entranced" the patient.[17]

Reason struck back when the king appointed two commissions to investigate these practices. Dr. Guillotin (of the blade) headed one group of four prominent doctors from the Faculty of Medicine. The other commission was headed by Ambassador Benjamin Franklin (of the lightning) and boasted five members of the Academy of Sciences including Bailly (of Jupiter) and Lavoisier (of oxygen). The commissioners spent weeks listening to Mesmeric theory and observing how its patients fell into fits and trances. They found false a report that being mesmerized through a door caused a woman patient to have a crisis. In Franklin's garden at Passy, a "sensitive" patient was led to each of five trees, one of which had been mesmerized. As the chap hugged each in turn to receive the vital fluid, he fainted at the foot of the wrong one. At Lavoisier's house, four normal cups of water were held before a mesmerized woman; the fourth cup produced convulsions, yet she calmly swallowed the mesmerized contents of a fifth, which she believed to be plain water. The commissioners concluded that there was no vital fluid: "the fluid without imagination is powerless, whereas imagination without the fluid can produce the effects of the fluid."[18] I'm reminded of Danny Kaye in *The Court Jester*: "The pellet with the poison's in the vessel with the pestle; the chalice from the palace has the brew that is true!"[19]

The verdict at Dover reminds us that the facts of evolution, no less than the laws of chemistry and physics, are the brew that is true.

Homeostasis and the East Wind

> I was attracted at one time toward philosophy. I recall walking home with Professor [William] James after one of his lectures and at the end of our talk confessing my inclination toward philosophical studies. He turned on me seriously and remarked "Don't do it, you will be filling your belly with east wind."
>
> —Walter B. Cannon[1]

The Physical and the Metaphysical

IN SEPTEMBER OF 1909, Sigmund Freud (age fifty-three) gave five lectures on "The Origin and Development of Psychoanalysis" at Clark University in Worcester. Together with Carl Jung and Sandor Ferenzci, he had been invited to America by Clark's President G. Stanley Hall to participate in a conference on modern aspects of psychology, and to receive the honorary degree of Doctor of Law. It was the only honorary degree Freud was ever awarded, and he recalled that "I found myself received by the foremost men as an equal . . . it seemed like the realization of some incredible day-dream: psychoanalysis was no longer a product of delusion, it had become a valuable part of reality."[2] Among those foremost men was William James M.D. (age sixty-seven), who had recently retired from Harvard, and had traveled from Cambridge for the occasion. James was impressed, but not entirely convinced, by the one lecture he heard. He wrote to his Swiss colleague, Theodor Flourney, a fortnight later:

> I went there for one day in order to see what Freud was like and met also Yung [*sic*] of Zürich . . . I hope that Freud and his pupils will push their ideas to their utmost limits, so that we may learn what they are [but] I can make nothing of his dream theories.[3]

Nevertheless, after posing for a group portrait outside the lecture hall, James and Freud took off for a *Spaziergang* around the rural campus. James had studied in Berlin and the two men conversed in easy German. Freud recalled that in the midst of their conversation James stopped suddenly, handed him a briefcase he was carrying, and asked Freud to continue walking, saying that he would catch up as soon as he had got through an attack of angina pectoris which was just coming on. James had been suffering from high blood pressure and bouts of shortness of breath for at least a decade. His attacks of angina were mounting.

Freud, the clinician, recognized that a diagnosis of increasing angina pectoris carried a glum prognosis at the dawn of the twentieth century. Freud knew that James was mortally ill and wrote sympathetically "He died of that disease a year later; and I have always wished that I might be as fearless as he was in the face of approaching death."[4]

And sure enough, despite several sessions with Christian Science healers, despite endless water cures at Bad Nauheim, William James succumbed to congestive heart failure at his summer home near Mount Washington, New Hampshire, in August of 1910. The autopsy, performed by his neighbor, Dr. George Shedd, showed an enlarged heart, and a dilated, arteriosclerotic aorta.[5]

James never wrote an account of his walk with Freud, but confided to Flourney that "Freud made on me personally the impression of a man obsessed with fixed ideas" and was disappointed that "Freud had condemned American religious therapy (which has such extensive results) as very 'dangerous' because so 'unscientific.' Bah!"[6]

That Jamesian "Bah!" is a verdict in keeping with the philosopher's growing attraction to a variety of religious and quasimedical experiences. An 1869 graduate of the Harvard Medical School, James's career had progressed at Harvard from a first appointment in Physiology, to directorship of the country's first Laboratory of Experimental Psychology, and finally to full flower as Professor of Philosophy in 1885. The titles of his major works reflect that path: a progression from the *The Principles of Psychology* (1890) to *The Will to Believe* (1897), via *The Varieties of Religious Experience* (1902) to *Pragmatism* (1907). He confessed "I always felt that the occupation of philosophizing was with me a valid excuse for neglecting laboratory work, since there is not time for both.[7]

James grew impatient with the messy details of lab science; he eventually became a vigorous opponent of vivisection. He was worried that modern science, wearing the blinkers of positivism, was learning more and more about less and less. He had warned an assembly of theologians at Princeton in 1881 against the experimentalists of the reflex arc,

> Certain of our positivists who keep chiming to us that amid the wreck of every other god and idol, one divinity still stands upright—that his name is Scientific Truth— . . . But they are deluded. They have simply chosen from among the entire set of propensities at their command those that were certain to construct, out of the materials given, the leanest, lowest, aridest result—namely the bare molecular world—and they have sacrificed the rest.[8]

James and Freud had both started their careers looking for Scientific Truth as chimed by giants of nineteenth-century biology. Both served apprenticeships in marine biology before entering into laboratories of experimental physiology. Having sidestepped the Civil War in 1865, William James was put to work sorting jellyfish on a taxonomic expedition to Brazil with Louis Agassiz. He next spent a *Wanderjahr* in Germany where in Berlin he learned experimental physiology first hand from Emil du Bois-Reymond, who had discovered electric currents in nerve and muscle. He was dazzled by Hermann von Helmholtz, who had not only formulated the law of the conservation of energy, but also invented physiological optics and invented the ophthalmoscope. James also studied the work of Helmholtz's young assistant, Wilhelm Wundt, who went on to establish the first laboratory of experimental psychology in Leipzig in 1875.[9]

Freud's science began in his student days with Carl Claus, a zoologist who sent Freud for a summer to the University of Vienna's *königliche und keiserliche* Marine Station at Trieste. Claus remains known today chiefly for having advised Ilya Metchnikoff that the *Fresszellen* the Russian had discovered should be called "phagocytes." Freud was set the painstaking task of determining whether eels had male gonads and where in the eel they were. After dissecting four hundred eels, he not only correctly located the testes of these creatures, a controversial issue at the time, but also material for his first publication.[10] Back in Vienna, Freud fell under the spell of Ernst von Brücke, the physiologist who had worked out major functions of the autonomic nervous system. In von Brücke's laboratory, Freud became a neurocytologist and, independently, went on to discover that cocaine is a powerful local anesthetic.[11] Von Brücke made it possible for Freud to study with Charcot in Paris—and the rest is the history of psychoanalysis.

The physiologists who taught James and Freud were positivist to the core. In 1847, Hermann Helmholtz, Emil du Bois-Reymond, Ernst von

Brücke, and Karl Ludwig (who invented the kymograph) formed the *Berliner Physicalische Gesellschaft* (Berlin Physical Society), popularly known as the "Helmholtz School of Medicine." Du Bois-Reymond outlined the group's philosophy in a letter to a friend in 1842:

> Brücke and I pledged a solemn oath to put into power this truth: no other forces than the common physical-chemical ones are active within the organism. In those cases, which cannot at the time be explained by these forces, one has either to find the specific way or form of their action by means of the physical-mathematical method, or to assume new forces equal in dignity to the chemical-physical forces inherent in matter, reducible to the forces of attraction and repulsion.[12]

The Berlin Physical Society argued that vitalism was dead and that all life processes take place in that bare molecular world which Freud hoped would one day contain his notion of mind. On the other hand, it is no accident that Boston was home to the Metaphysical Club, formed in 1872 by William James, Oliver Wendell Holmes, Jr., and Charles Peirce. The postbellum city was recovering from the twin onslaughts of the Civil War and Darwinian theory;[13] few on this side of the Atlantic and certainly fewer at Harvard were prepared to accept European reductionism à la Helmholtz or Hippolyte Taine (see "Reducing the Genome").

One might say that William James spent his life pursuing those "new forces equal in dignity to the chemical-physical forces inherent in matter" that duBois-Reymond dismissed. Suffering from one somatic ache or melancholic spell after another, James had over the years consulted practitioners of homeopathy, balneotherapy, mesmerism, and spiritualism. He was a staunch defender of Christian Science, a doctrine of American spiritual healing that arose *de novo* in the Boston of 1875. On its behalf James urged the Massachusetts legislature to table a law that would require a medical degree for the treatment of diseases of the nervous system and to "resist to the uttermost any legislation that would make 'examinable' information the root of medical virtue . . ."[14] James's interest in the nonexaminable led him to study the fashionable clairvoyants of his day and to found the American Society for Psychical Research. Over three decades he recorded the séances of a Mrs. William J. Piper of Boston, whom he recommended to his skeptical sister as one " in possession of a power as yet unexplained."[15] East wind, indeed!

Shortly before Freud mocked American credulity at Clark, James had attended a séance with another medium, a Madame Eusapia Paladino, of

Naples, in the course of which James experienced a "Queer twisting of my chair." His less credulous colleague at Harvard, Josiah Royce, teased James with the jingle:

> Eeny, meeny, miney, mo,
> Catch Eusapia by the toe,
> If she hollers, then you know
> James's theory is not so.[16]

Undeterred, William James published a hundred-page report of his communications with the deceased Richard Hodgson, in July of 1909. His last effort in the area, an essay entitled "The Final Impressions of a Psychical Researcher," was written in October 1909 and survives today on a cranky Web site.[17] On the occasion of James's death and years after their youth in the Metaphysical Club, Oliver Wendell Holmes, Jr., explained to a colleague why their ways had parted: "distance and other circumstance and latterly my little sympathy with his demi-spiritualism and pragmatism were sufficient cause. His reason made him skeptical and his wishes led him to turn down the light so as to give miracle a chance."[18] Freud, the psychiatrist, had spelled out a different view of liaisons with the psychic realm:

> In point of fact I believe that a large part of the mythological view of the world, which extends a long way into the most modern religions, is nothing but psychology projected into the external world. The obscure recognition . . . of psychical factors and relations in the unconscious is mirrored—it is difficult to express it in other terms, and here the analogy with paranoia must come to our aid—in the construction of a supernatural reality, which is destined to be changed back once more by science into the psychology of the unconscious.[19]

These days we tend to doubt that a psychology of the unconscious can explain beliefs as dotty as clairvoyance or as persistent as American spiritual healing. But surely Dr. Freud, the clinician, had it half right: American spiritual healing wasn't very helpful for diseases like angina pectoris. On the other hand, conventional medical treatments of the day, including those baths at Bad Nauheim were no more effective. It's taken medicine over a century to do a better job. Nowadays we have a fair understanding, at the nutritional, psychosocial, and molecular level of how angina pectoris comes about and what causes it: it's a clutching pain in an anxious chest due to spasm of coronary arteries narrowed and inflamed by deposits of fatty sludge. We've gained that understanding by following principles laid down by the Helmholtz school of medicine:

> physiologists must [pledge] unconditional conformity to the laws of nature in their inquiries . . . they will have to apply themselves to the investigation of the physical and chemical processes going on within the organism.[20]

Hallelujah for Homeostasis

If, in the twenty-first century, we're better at the "investigation of physical and chemical processes" within the human organism, much of that progress is due to a student of William James, Walter B. Cannon (1871–1945). Cannon discovered that one substance in that vast, bare, molecular world can both stir the emotions and trigger angina. It's adrenaline, or epinephrine, isolated in 1896 by an American student of Carl Ludwig, John Jacob Abel. Nowadays blockers of its beta-receptors are among the drugs most widely used to treat heart disease—and anxiety.

On January 16, 1911, half a year after James's death, Cannon exulted in his notebook: "This is hallelujah day! De la Paz and I get clear evidence of emotional production of adrenalin in cat."[21] Working in the Harvard physiology laboratories, Cannon and his assistant had skillfully placed a catheter into the vena cava of a cat under local, cocaine anesthesia (à la Freud) and sampled blood coming from the adrenal vein both before and after a dog barked in the same room and then compared "quiet" against "excited blood." As a bioassay they took advantage of the capacity of adrenaline to relax a strip of smooth muscle. Recording these events on a kymograph (à la Ludwig), they found that "excited blood" contained an active substance that raised blood pressure, elevated blood sugar, made sugar spill over into urine, dilated the pupils, and made hair stand on end. Four days later Cannon confided to his notebook:

> Got idea that adrenals in excitement serve to affect muscular power and mobilize sugar for muscular use—thus in wild states readiness for fight or run![22]

Cannon and de la Paz went on to show that "Injected epinephrine is capable of inducing an atheromatous condition of the arterial wall in rabbits" and that athletes competing in the Harvard/Yale boat race had dramatic rises in their blood sugar. The experiments were rapidly published and aroused wide public interest.[23] Since Cannon and de la Paz had produced a transient state of diabetes and mimicked atherosclerosis in animals, they were encouraged to summarize their work in the *Journal of the American Medical Association*. Their summary concluded that "some phases of these pathological states are associated with the strenuous and exciting character of modern life acting though the adrenal glands."[24] They had also reduced a complex emotional response, "fright," to a secretion of the glands. It's an idea as old as the Enlightenment. Denis Diderot had, in fact, anticipated Walter B. Cannon's notion of fight or run in his *Réve d'Alembert*:

> In fits of strong passion, in delirium or at times of imminent danger, if the master throws all the forces under his command towards a single objective, then even the weakest animal develops incredible strength.[25]

Claude Bernard, the Carl Ludwig of France (or vice versa) extended the principle in the 1850s, formulating the notion of a "milieu internal," that constant, steady state of our bodily fluids which we disturb at our peril:

> The organism tends to maintain itself and the nervous system alerts the organism and it reacts in consequence. If the environment changes, the organism tends to maintain itself in the new environment, it acclimatizes, it yields, etc.[26]

The tone of the Gallic passage is more accommodating than the dramatic image of fight or flight on the Charles River: but then the Collège de France never had a heavyweight crew. Cannon broadened his observations on the body's reaction to adrenaline in 1926, when he formulated one of the more holistic concepts in biology, that of "homeostasis," a dynamic extension of Claude Bernard's theory of the internal milieu. Cannon explained that the Gallic balance of internal fluids, the constancy of our blood, sweat, and tears was maintained by a series of self-regulating pressure/volume circuits under the control of hormones. Those circuits maintain not only the constancy of inner fluids, but permit our body as a whole to adjust to change:

> The steady states of the fluid matrix of the body are commonly preserved by physiological reactions . . . Special designations are therefore appropriate: homeostasis to designate stability of the organism; homeostatic conditions to indicate details of the stability.[27]

Cannon announced homeostasis in a volume dedicated to the most holistically inclined of William James's colleagues, the Parisian Professor of Physiology, Charles Robert Richet. Richet (1850–1935) was an experimental physiologist who, *sans évidence expérimentale,* anticipated Cannon. In 1900, Richet suggested that organisms remain stable because they adapt, and that "a slight instability is the necessary condition for the true stability of the organism."[28] He won a Nobel Prize in 1913 for his studies of anaphylaxis—the kind of immediate, immune paroxysm induced by a bee sting, for example, a reaction which can be quelled by adrenaline. Richet was also an early amateur aviator, a novelist, a playwright, and—like William James—a sometime president of the Society for Psychical Research. He truly believed in telepathy, in telekinesis, and in the reality of ectoplasmic phenomena. He was a high-flying avatar of the Jamesian will to believe. How fitting that William James composed his "The Varieties of Religious Experience" in the spring of 1900 while James and his wife were guests at Richet's luxurious Château Carqueirianne on the Riviera.

Cannon, the American Claude Bernard

Cannon himself was no slouch as a writer. Shortly after he satisfied himself that the adrenals mediated fight or flight, he wrote a book of general appeal: *Bodily changes in pain, hunger, fear, and rage; an account of recent researches into the function of emotional excitement.*[29] He went on to produce an account of homeostasis, *The Wisdom of the Body*,[30] that found a wide audience and became required reading for a generation of medical students. His lyrical memoir, *The Way of an Investigator*, forged a link in that chain of Harvard Medical School physician/scientists who were also lucid writers.[31] Their lives spanned almost two centuries from 1809 to 1993. Dr. Oliver Wendell Holmes taught William James who taught Walter B. Cannon who taught Lewis Thomas. It seems appropriate, therefore, that Lewis Thomas's major discovery, the case of the floppy-eared rabbits, was called "serendipitous." Cannon had introduced Thomas—and the biomedical community at large—to Hugh Walpole's Serendip (the old name for Ceylon) in his memoir:

> [Serendipity] is said to designate the happy faculty, or luck, of finding unforeseen evidence of one's ideas or, with surprise, coming upon new objects or relations that were not being sought.[32]

It also seems appropriate that Cannon had written his undergraduate thesis at Harvard (1894) on Oliver Wendell Holmes and—in the heat of enthusiasm—gushed that: "[Holmes's] lecture on homeopathy is conceived and written in a vein of noble scorn and the thought is poured out along the pages with a lucidity, pungency, and satire, and cogent understatement that gives the performance the velocity of a cannon shot."[33]

Cannon shot, indeed! The young senior could have been describing himself. At Harvard, he was a poor kid from the Middle West amongst the sons of the Puritans; but he soon found a home in science. He entered Harvard in 1892 at time when Eliot's curriculum reform had provided research opportunities aplenty for undergraduates. As had James and Freud before him, Cannon began science as a marine biologist; Alexander Agassiz (Louis's son) gave Cannon his first summer research job at his private Newport laboratory, and when Cannon returned to Cambridge, he signed up for James's experimental psychology seminar. William James had brought over Hugo Münsterberg, a student of Wundt's, to direct the daily laboratory work in psychology; it was Münsterberg who recommended a bright Radcliffe undergraduate, Gertrude Stein, for the James seminar in 1893. Stein recollected that one student was working on notions of religion, another on chicken development, a third, Leon Solomons, was experimenting with automatic writing. Stein joined Solomons and her first publication on motor automatism came from that work.[34] Münsterberg, as much as James, attracted Cannon to experimental science. Cannon went on to earn all the honors at Harvard College: he was elected junior Phi Beta Kappa, and graduated summa cum laude in 1896. Torn between a career in medicine and one in philosophy, Cannon followed William James's advice, he scorned the East wind, and became America's Claude Bernard.

Cannon was perhaps the most prominent American medical scientist of his generation trained entirely in the United States; indeed, formal training in physiology did not begin at Harvard until the late 1870's. In Paris, François Magendie (Bernard's teacher) had founded the *Journal de physiologie expérimentale* in 1821, and in 1833 Berlin established the first chair of physiology for Johannes Müller (who taught Helmholtz, Ludwig *et al.*). President Eliot made up for lost time. He plucked Henry Bowditch, scion of a distinguished Boston medical family, from Carl Ludwig's lab in Leipzig to make him the first full-time faculty member hired at the Harvard Medical School. Almost immediately as Cannon entered the Medical School, Bowditch became Walter Cannon's mentor, guide, and supporter: "I am the grandson of Ludwig in scientific inheritance and the son of Bowditch" Cannon announced proudly on a later occasion.[35] The Helmholtz school of medicine had crossed the Atlantic.

In consequence, instead of filling his belly with east wind, Cannon filled it with barium. Elaborating on studies made in Bowditch's department as a medical student, he made the pioneering discovery that one could visualize the human gastrointestinal tract by means of radio-opaque substances such as barium. He went on to show that emotions could cause changes in the tone and function of the GI tract and these discoveries put his reputation on the map. Then came his great find: Cannon's description of the alarm reaction was a physiologist's account of how Darwinian evolution might play itself out in the field:

> The organism which with the aid of increased adrenal secretion can best muster its energies, can best call forth sugar to supply the laboring muscles, can best lessen fatigue, and can best send blood to the parts essential in the run or the fight for life, is most likely to survive.[35]

Social Homeostasis

Cast into rhyme as the "flight or fight" reaction, homeostasis was validated in the battalion aid stations and field hospitals of the Great War; the face of fear was drawn by adrenaline. It's probably no coincidence that the terms "allergy" (altered action), coined by von Pirquet in the Vienna of 1906, and "anaphylaxis" (a shield turned against itself), coined by Richet in the Paris of 1901, entered medical literature as the industrialized nations of Europe mobilized for the fight for life of the Great War. Once again, allegiance to the nation-state split the liberal fraternity of science; German optics gave us the periscope, British chemistry gave us dry explosives. Dreaming of empires to rival Rome, the youth of Europe learned the lesson of Horace:

> *Angustam amice pauperiem pati*
> *Robustus acri militia puer*
> *Condiscat et Parthos ferocis*
> *Vexet eques metuendus hasta . . .*
> *Dulce et decorum est pro patria mori . . .*

> (A stint in the army will shape up our young
> they'll learn to crave danger and conquer defeat,
> learn how our Roman, lance-wielding horsemen
> force the fierce Turk to flee in retreat...
> How sweet and how glorious to die for one's country!)[36]

As Richet suggested, when the shield of antibody (*philaxos*) becomes inverted (*ana*), defense can become self-destructive; the Guns of August proved him right. Both William James and Walter B. Cannon saw war coming, but offered somewhat different solutions. In lieu of a stint in the army, James proposed "The Moral Equivalent of War" to shape up the young: a year or two of public service in a construction project, a job in industry, or ministry to the poor (*i.e.*, a pre-mature CCC, or Peace Corps).[37] Having found that Harvard football players spilled sugar in their urine in response to the adrenaline rush of the big game, Cannon suggested "The Physiologic Equivalent of War."[38] He was sure that Olympic-style contests would satisfy that ancient drive to force the fierce Turk to flee in retreat; or as Orwell had it "war without shooting." Neither suggestion averted the Great War: it remained for Wilfred Owen to warn war lovers that were they to see the wounded:

> My friend, you would not tell with such high zest
> To children ardent for some desperate glory,
> The old Lie: Dulce et decorum est
> Pro patria mori.[39]

Cannon served with great distinction in Wilson's "War to Make the World Safe for Democracy," witnessing the bloodshed of battalion aid stations and working on experimental shock in London and Paris. He progressed in rank from lieutenant to lieutenant colonel, and earned honors from England, France and the United States for his contributions to military medicine. His book *Traumatic Shock* spelled out the lessons learned in battle and in the lab: traumatic shock is due to loss of fluid and blood from dilated, leaky capillaries, a process he called "exemia." The treatment of shock, he correctly deduced, was to replace fluid and to maintain the circulation.[40] It's the same lesson I was taught as a young doctor at Fort Sam Houston two wars later.

After the Great War, Cannon carried his concept of homeostasis into the social realm. He was a political progressive, and devoted much of his energy in the late 1930s to anti-fascist causes. The poor kid from the Midwest became a voice for social democracy at home and abroad. Cannon had been a good friend of Dr. Juan Negrin, Professor of Physiology in Madrid and sometime premier of the Spanish Republic. As a "premature anti-fascist," Cannon was proud of his support for "republican forces in Spain and in China as they struggled against oppression."[41] Beginning in 1936 he served for two years as national chairman of the Medical Bureau to Aid Spanish Democracy. The group sent to Spain medical personnel, surgical instruments, hospital equipment, and ambulances amounting in value to more than a million dollars. "Naturally I was charged with being a Bolshevik . . . an enemy of the Catholic Church, and in general a Red . . ."[42] In keeping with his Popular Front politics he proposed that the concept of homeostasis, properly modified, could be used to explain the human need for material and scientific progress. "The functioning of the human brain has made social homeostasis differ markedly from physiological homeostasis . . . An upset of constancy necessarily results."[43] Social homeostasis, as proposed by Cannon, required evolution of the body politic, not the blind support of received truth:

> Man did not fall from a state of grace and therefore does not require to be redeemed; instead through a prodigious process of evolution, he has gradually risen, in spite of frequent backslidings, from the brutish state to higher and higher degrees of civilized development.[44]

In the darkest days of the twentieth century Cannon was confident that the "Wisdom of the Body" had social significance.

Cannon, the anti-Fascist

When Hitler occupied Austria in March of 1938, the cruel repression by Austrian authorities of Jewish physicians and scientists made it imperative for them to flee. Opportunities for posts abroad were few and far between. Indeed there was wide opposition both in England and the United States to a sudden, unwelcome influx of alien professionals. Cannon played an honorable role in stemming the nativist tide. In Massachusetts, a bill was submitted to a committee of the legislature in November of 1938 that would require full U.S. citizenship for practice in the Commonwealth. But, another committee was also at work in Boston. One mission of the Committee to Aid Émigré Physicians was to help relocate Jewish physicians and scientists in the United States; among its members were Dean David Edsall, Tracy Putnam, Oliver Cope, and Jacob Fine of the Harvard Medical School. This committee prevailed on Cannon, by virtue of his eminence and political experience, to lobby the chairman of the legislature, a C. C. Fuld, to table the bill. Cannon lobbied mightily, and the bill never reached the floor.[45] Harvard's debt to Helmholtz might be said to have been paid a hundredfold.

Two distinguished scientists owe their American escape to Cannon's personal effort. Ernst von Brücke, Professor of Physiology at Innsbruck and grandson of Freud's mentor, was under direct threat of death; his mother and wife were Jewish. Cannon arranged for his colleague at Harvard, Alexander Forbes, to guarantee financial support for a research fellowship for von Brücke. When the fellowship did not prove enough to secure a visa for von Brücke and family, Forbes and Cannon persuaded President Conant to offer a lectureship, and urged his friends to join in contributing the needed funds. After von Brücke was forced to sell his home in Innsbruck and pay $45,000 in ransom to the Nazis, the family arrived in Boston in August of 1939, two weeks before Germany invaded Poland.[45]

Otto Loewi, Professor of Pharmacology at Graz, had won the Nobel Prize in 1926 for his discovery of acetylcholine, a.k.a *Vaguststoff*. No matter how high the prize, when the Nazis seized Austria in 1938, he was awakened from his sleep by a dozen young storm troopers, armed with guns, who hustled him to jail:

> When I was awakened that night and saw the pistols directed at me, I expected, of course, that I would be murdered. From then on during days and sleepless nights I was obsessed by an idea that this might happen to me before I could publish my last experiments. After repeated requests, a few days later I was permitted to have a postal card and a lead pencil to write, in the presence of my guard, a communication of my last experiment to be sent to *Die Naturwissenschaften*.[46]

Loewi owed his release from prison and an uncertain refuge in London to the co-holder of his 1936 Prize, Sir Henry Dale. Loewi had been compelled to instruct the Swedish bank in Stockholm to transfer the Nobel Prize money to a prescribed Nazi-controlled bank.[47] Once safe in England, Loewi applied to Cannon for a position at Harvard. By then, an influx of the great and near-great of European science at Harvard had made the university reluctant to offer even a Nobel laureate a faculty position, but Cannon found another solution. He persuaded Homer Smith, the Professor of Physiology at the NYU School of Medicine to offer Loewi a full-time professorship in pharmacology; Smith had been a research fellow in Cannon's lab in 1925–1926. Loewi arrived for work in New York in 1940, and spent the rest of his life teaching the joy of discovery at NYU in the winters and at the Marine Biological Laboratory in Woods Hole in the summers. He is remembered fondly in both places.

In the late 1950s, Loewi was sitting before the old shingled mess hall at Woods Hole when a young MBL scientist rushed up to him exuberantly:

"I've got it! I know why the squid axon requires calcium ions!"

"Congratulations!" said Loewi, "You will be right, for a while."[48]

It is no accident that Cannon's autobiographical *The Way of an Investigator* began as a series of lectures he delivered at NYU in the first years of America's war with Hitler. Homer Smith and Otto Loewi had arranged for Cannon to come to NYU after his mandatory retirement from Harvard, and Cannon was happy to be "appearing vertical again before a roomful of medical students. . . ."[49] And it is no accident either that on a wall of our cottage at Woods Hole hangs my father's license to practice medicine. It was issued on March 14, 1940, by the Commonwealth of Massachusetts to Dr. Adolf Weissmann, a graduate of the University of Vienna, an émigré who had arrived in America barely a year and a half before. Cannon had made that license in Massachusetts possible.

Relationships of this sort may or may not have been in Cannon's mind when he worked out his progressive notion of social homeostasis, but family is family, especially in science. When Cannon acknowledged that he was the grandson of Ludwig and the son of Bowditch, he was equally proud that he was now "father of a scattered brood [and] some of my sons have sons themselves, and I am therefore a proud grandfather."[49] Since I count Homer Smith, Otto Loewi, and Lewis Thomas among my teachers, I remain pleased that my distant scientific ancestor, Walter B. Cannon, chose to abjure the east wind and stick to experimental science.

Red Wine, Ortolans, and Chondroitin Sulfate

> Yves Mirande would dazzle his juniors, French and American, by dispatching a lunch of raw Bayonne ham and fresh figs, a hot sausage in crust, spindles of filleted pike in a rich rose sauce Nantua, a leg of lamb larded with anchovies, artichokes on a pedestal of foie gras, and four or five kinds of cheese, with a good bottle of Bordeaux and one of champagne, after which he would call for the Armagnac and remind Madame to have ready for dinner the larks and ortolans she had promised him, with a few langoustes and a turbot. . . .
>
> —A. J. Liebling, *Between Meals*[1]

> At the age of twenty I believed that the first duty of a wine was to be red, the second that it should be Burgundy. . . . I have lost faith in much, but not in that.
>
> —Alec Waugh, *In Praise of Wine and Certain Noble Spirits*[2]

AMERICANS ARE SOLD CARLOADS OF "food supplements" that are useless and taste bad; the French supplement their food with wine that is wholesome and tastes good. Americans are upset that the French gavage their geese to make *foie gras*, the French are upset that Americans gavage their prisoners at Guantanamo. One notes that in the life expectancy rankings of the UN, France ranks 10th in the world (men and women combined=79 years) while the US clocks in at 19th (77 yrs).[3]

One is forced to conclude that gourmandes like playwright Yves Mirande and novelist Alec Waugh had it right: Mirande died in 1957 at age 82, Waugh died in 1981 at age 83. The ... nd
red wine, living to a ripe old ... ne
"French paradox." The traditi... is
death from prostate cancer, ... is
Mitterand, gave a terminal dinn... s
a "feast of ortolans." He was in

The ortolan, a small darling ... e
the days of Brillat-Savarin (d. ...)

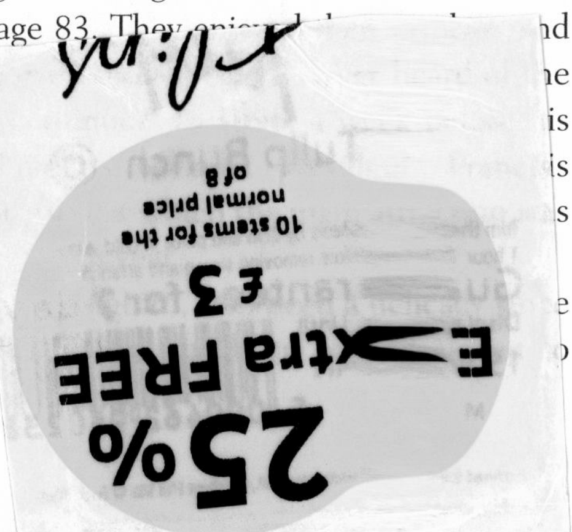

extinction. But while French law prohibits netting and eating of the little creatures, they are still available to the low and the mighty. The bird is netted during its autumnal migration, fattened up for a few days in dark captivity, and drowned alive in jugs of Armagnac. It is then roasted to sizzling, decapitated, and sucked up in one bite, bones and all, through its rectum. Tradition has it, and Mitterand followed suit, that ortolans are eaten under a napkin that covers both dish and diner, to enhance the aroma—or to cover the sin.[4] Since ortolans are best served with red wine it surprised few that when the contents of Mitterand's cellar were auctioned at Drouot last year, its unopened treasures fetched over $17,000.[5] However heartless the feast of ortolans, M. le Président was free of heart disease on his death. American statesmen who shoot quail with buckshot after beery barbecues cannot claim clean coronaries.

Since the 1980s, it's been appreciated that the French take in as much cholesterol and saturated fat as Brits, Scandinavians, or Americans, but 40 percent fewer Frenchmen die of cardiovascular disease. Their lower mortality has been attributed in great part to wine consumption. Indeed, Serge Renaud, who first framed the French paradox,[6] studied 34,000 middle-aged Frenchmen and concluded that the cardiovascular mortality of those who consumed a mean of two glasses of wine a day (48 g of alcohol) was lower by over 30 percent than that of French abstainers. The abstainers died at the rate of Brits or Americans; neither blood lipids nor other risk factors accounted for the protection afforded by (mostly red) wine.[7]

This year, the dietary habits of Mssrs. Mirande and Waugh, red wine with figs and foie gras, have gotten better notices than such American manipulations as fat restriction, calcium supplements, or extracts of cow cartilage (chondroitin sulfate.) Large, long, broad-based, US government-supported studies have recently shown—in headlines taken from the *New York Times*—that a Low-Fat Diet Does Not Cut Health Risks, that there is No Clear Benefit Of Calcium Pills for postmenopausal bone fractures, and that the dietary "supplements" glucosamine and chondroitin sulfate, wrongly billed as drugs, don't work in osteoarthritis: Top-Selling Arthritis Drugs Are Found to Be Ineffective. (Each of these studies falls into the "*et al., ad infinitum*" category, i.e., a massive, longitudinal study with more signatories than the Declaration of Independence.)

In the chondroitin sulfate/glucosamine study, 60 percent of the osteoarthritis patients given placebo got better, those given supplements fared no better, but a Cox 2 inhibitor (celecoxibid) proved effective. Effective, that is, by the less than stringent criteria used in rheumatology (20 percent relief of arthritic symptoms).[8] Would that the joint docs in America were as rigorous as the French heart doctors, who used a more robust end point to show the efficacy of fermented grapes as a food "supplement": a 40 percent reduction of mortality!

Sabbaticals in France have persuaded me that the French view food as a contribution to the sum of earthly pleasure, which requires no other supplement but wine. In America, where hirsute wellness gurus peddle cartilage powder in unpalatable pills, food is considered a duty paid on health: no French waiter has ever asked me "Are you still working on it?" We tend to view meals as rituals performed for the sake of our bodies: like push-ups at daybreak. Years before exercise waifs persuaded adults to consume power bars and Aquafina, Evelyn Waugh (Alec's brother) cabled back to Ian Fleming from a Hollywood studio: "The men lunch in wine-less canteens." We've never recovered from our heritage of Graham crackers and prohibition. A recent University of Pennsylvania study concluded that of several countries studied, we Americans, who do the most to alter our diet, are the least likely to classify ourselves as healthy eaters. The US group associated food most with health and least with pleasure, while the French group, which was the least health oriented, ranked highest on the food-pleasure scale.[9] Brillat-Savarin had the verdict on this: *"La destinée des nations dépend de la manière dont elles se nourrissent."* ["Where the country goes depends on how it is fed."][10]

Meanwhile, red wine, and its second most interesting bioactive ingredient, resveratrol, have gained kudos as wonder drugs. The media, pleased that resveratrol may be an explanation for the French paradox, have reported, not incorrectly in my view, that the compound shows promise in conditions that range from arthritis to cancer: NOW RED WINE CAN EASE YOUR SORE JOINTS (*Daily Mail*, London, 2005). WINE BEATS ALZHEIMER'S (*Evening Times*, Glasgow, 2005). RED WINE FOR ROBUST PROSTATES (*New York Times*, 2004). Waugh and Mirande had it right: CHEMICAL ABUNDANT IN RED WINE APPEARS TO SLOW AGING (*Boston Globe*, 2003). Nor has resveratrol escaped notice in the scientific literature, chalking up 1,396 articles in the Pub. Med. database since 1985; including a recent review in the journal for which I am responsible.[11]

Resveratrol, otherwise known as trans-3,4'5-trihydroxystilbene, is used by the grapevine as a defense against fungal invasion, becomes concentrated in grape skin and is largely absent from white wine, beer, and whisky. It has profound effects in many biological systems: at concentrations present in the blood after two glasses of red wine, it prevents clumping of platelets and white cells. At similar concentrations, resveratrol prolongs the lifespans of yeast, flies, fish, and roundworms, where its antiaging effects have been pinned down to a family of genes that mediate responses to stress.[12] Resveratrol also acts as a plant-derived estrogen, with many of the biological effects of natural estrogens.[13] It is related structurally to diethylstilbesterol, another female hormone surrogate, and to hyroxystibamidine, a venerable agent active in parasitic diseases and models of inflammation.[14] The stuff even clears out the tangled debris deposited in Alzheimer's disease.[15] In the cells of mice and men, resveratrol blocks activation of a key regulator that turns on the genetic machinery common to both tumor induction and inflammation. Indeed, its site of action in this mode is the very same site that is inhibited by antiinflammatory levels of aspirin-like drugs.[16] Resveratrol inhibits prostaglandin synthesis via the COX-1 / COX 2 pathway. Indeed, resveratrol at concentrations that can be attained after two glasses of red wine is as active an inhibitor of cyclooxygenases as those antiarthritis drugs that beat out chondroitin sulfate.[17]

So, the next time you read about the results of one of those *et al., ad infinitum* studies about the effects of food supplements such as cow cartilage extracts on arthritis or cancer, ask yourself if the results are better than those obtained with two glasses of red wine—or aspirin, for that matter.[18] In fact, it's probably no accident that resveratrol from a grapevine in France acts at the same intracellular target as salicylate from a willow in England or a COX inhibitor from the lab: living things, be they animal or vegetable, make their living on the same planet. One creature's need is another one's supplement. That's neither coincidence nor intelligent design—it's evolution.

Cortisone and the Burning Cross

RHEUMATOLOGY, THE TREATMENT OF bones and joints and widespread miseries, came late to the game of medical science. For many years my medical specialty was a descriptive art; in any meaningful way we had no idea of what was going on. The heart doctors had their cardiograms and digitalis, the endocrine people had their thyroid tests and extracts, but joint doctors seemed condemned to stand idly by to watch their patients turn into cripples after one or another stopgap treatments. Oh yes, we had diathermy, gold salts, paraffin injections and, believe it or not, bee venom. We knew how to treat gout with colchicine, learned to give penicillin to prevent rheumatic fever, but by and large our treatment of joint disease, or serious threats like systemic lupus erythematosus (SLE) was limited to aspirin, aspirin, and more aspirin. All that changed in the *annus mirabilus* of our field, 1948. It's the year that cortisone was first given to a patient with arthritis. It's also a year when bigots were burning the houses of black folk in white suburbia and lighting crosses on their lawns.

At a staff meeting of the Mayo Clinic in January of 1948, Malcolm M. Hargraves described a strange kind of cell that formed in blood samples of patients with SLE. The disease, which tends to afflict young women, attacks joints, skin, kidney, heart, and brain. Before 1950, we couldn't really tell who had SLE and who didn't; we had no clue as why it was so often fatal. Hargraves had discovered what he called the LE cell, which finally permitted us not only to make a diagnosis of the disease, but also told us what was going wrong with these poor women. The LE cell, it turned out over the years, is a white blood cell (a neutrophil) that has ingested the dying nucleus of another cell against which lupus patients make antibodies. It also turned out that those antibodies against the nucleus and/or its constituents—the anti-DNA antibodies—were just the tip of an iceberg. SLE patients make a dazzling number of antibodies against bits and pieces of their own cells. Their immune system recognizes such bits of "self" as if they were a microbe, a tad of "nonself" that wants expunging. Hargrave's discovery of the LE cell sparked the study of autoimmunity and lifted rheumatology over the threshold of science.[1]

In the same month, immunologist Harry Rose and rheumatologist Charles Ragan of Columbia described a factor in the serum of most patients with rheumatoid arthritis (RA) that clumped sheep red blood cells coated with human antibodies: the "sensitized sheep cell agglutination test." Tests for this factor not only permitted accurate diagnosis of rheumatoid arthritis, but also taught us how joints are attacked in RA. What came to be called "rheumatoid factor" turned out to be yet another autoantibody, of great size and with a tendency to form sludge in the blood. Normal human antibodies, the "self" in this case, were recognized as "nonself" by rheumatoid factor. The agglutination reaction in a test tube was a pretty good reflection of what happens in life. In patients with RA, blobs of antibodies containing rheumatoid factor form in the blood like *îles flottant*; they become trapped in joint spaces, joint cells try to get rid of the unwanted debris, cry havoc, and let loose the dogs of inflammation. As with Hargraves and the LE cell, the discovery of rheumatoid factor made it possible to make sense of yet one more of our diseases.[2]

On April 20, 1949, William A. Laurence of the *New York Times* broke news of another discovery anounced at a staff meeting at the Mayo Clinic:

> Preliminary tests during the last seven months at the Mayo Clinic with a hormone from the skin of the adrenal glands has opened up an entirely new approach to the treatment of rheumatoid arthritis, the most painful form of arthritis, that cripples millions, it was revealed here tonight.[3]

That evening, Philip Hench, Charles Slocumb, and Howard Polley reported their experience with fourteen cases of rheumatoid arthritis treated with a precious material called "Kendall's compound E" or 17 hydroxy-11 dehydrocorticosterone.[4] Cortisone had entered the clinic.

Within a week, cinemas nationwide showed newsreels of cripples rising miraculously from their wheelchairs. By May of 1949, Hench and coworkers reported the "complete remission of acute signs and symptoms of rheumatoid inflammation" at the Association of American Physicians in Atlantic City. In June they added success with rheumatic fever to the cortisone legend at the Seventh International Congress of

Rheumatic Diseases in New York. It was the summer I decided to follow my father into rheumatology and, in retrospect, I'd guess that it was cortisone that convinced me. I will never forget the waves of applause after Hench's dramatic film clips were shown to a packed crowd at the Waldorf-Astoria. One could actually *do* something about a crippling disease like rheumatoid arthritis.

In October 1950, the Nobel committee announced that Philip Hench and the two biochemists who had painstakingly isolated and described the chemistry of adrenal steroids, Thaddeus Reichstein (University of Basel) and Edward Kendall (Mayo Clinic), would receive the Nobel Prize in Physiology or Medicine for "their discoveries relating to the hormones of the adrenal cortex, their structure and biological effects."[5] Hench remains the only rheumatologist among Nobel laureates. So universal was the acclaim for cortisone that the Swedish announcement of the 1950 Nobel Prize in literature (William Faulkner) was almost a footnote in the world press.

There were other footnotes in the fall of 1950. On Thanksgiving eve, November 22, a hate crime was committed in the exclusive Oak Park suburb of Chicago: ARSON FAILS AT HOME OF A NEGRO SCIENTIST headlined the newspapers.[6] It was one of a string of cross-burnings and arson attempts in the white suburbs of Chicago. The scientist in question was Percy Lavon Julian (1899–1975), the first African-American to buy a home in Oak Park. Julian was described as "president of soybean research at the Glidden Corporation, discoverer of life stimulating [*sic*] chemicals and drugs for treatment of diseases."[6] More to the point, in November of 1950 Julian was working feverishly on the practical synthesis of cortisone via Reichstein's compound S, work that resulted in US patent #2,752,339, "Preparation of Cortisone."[7]

On December 11, 1950, less than a month after Julian's house was torched in Oak Park, Hench addressed the Nobel audience at the Karolinska Institutet.[8] He rehearsed the long trail of his discovery: how in the twenties he had first noted relief of rheumatoid arthritis in a male physician who developed jaundice; how in the thirties he had noted that pregnancy relieved the disease in female patients; how in the forties he had discussed with Kendall the possibility that substance X in the blood of jaundiced or pregnant patients might be his compound E; finally, how in September of 1948 he had written to Merck for small amounts of the laboriously synthesized material to test in the clinic. His letter noted that jaundice or pregnancy brought almost immediate relief; he promised Merck that "if any adrenal compound is of real significance in rheumatoid arthritis we would expect to see some results within a very few days." Three days to be exact. Beginning at 100 mg/day, given intramuscularly, the Mayo doctors obtained dramatic results; they soon lowered the dose to a "maintenance" dose of 25mg of cortisone, Hench and Kendall's new name for compound E. That's equivalent to 25mg tapering to 5mg of prednisone and nowadays those results are duplicated daily the world over.

In his Nobel speech, Hench reminded his audience how difficult it was to manufacture practical amounts of cortisone. Merck had gotten into the steroid business during World War II when the National Research Council subsidized a crash program for synthesis of adrenal steroids. Washington had learned that Luftwaffe doctors were experimenting with injections of adrenal extracts to keep their aviators stress-resistant at 40,000 feet, and several of Kendall's compounds (E and F especially) seemed likely candidates. Merck's Lewis Sarret came up with a complex, difficult synthesis of E from bile: by 1944 it had produced 15 milligrams from the bile of 2,500 cows![9] Hench averred that "although none of the thirty-six steps required to convert desoxycholic acid into cortisone has been by-passed, some of the steps have been made less costly, less time consuming, and productive of greater yields." His fellow laureate, Thaddeus Reichstein told his Stockholm audience that "For practical purposes [the Sarrett] method is much too laborious. In the last two years, again particularly in the U.S.A., at the cost of a considerable amount of time, much better methods have been discovered [e.g. Julian and his collaborators . . .] For after the clinical results of Hench, Kendall and their colleagues it can hardly be doubted that the future demand for these substances will be very great."

"Future demand" was met as the cost of production of cortisone fell from $1000/gm in 1948 to $150 in 1950 to less than $7 in 2000.[10] We owe this boon to the synthesis of cortisone from vegetal sources by Percy Lavon Julian, that brilliant "Negro scientist" whose house in Oak Park was torched on Thanksgiving eve of 1950.

Percy Lavon Julian of Montgomery, Alabama, was the grandson of a former slave and son of a postal employee.[11] He worked his way through DePauw University waiting on tables and graduated as class valedictorian with a Phi Beta Kappa key. After DePauw, he served teaching stints at several black colleges and finally received a fellowship to Harvard where he earned an MA in chemistry. Since Harvard in the 1920s had no place for a black scientist, Julian applied—successfully—for a Rockefeller Foundation fellowship at the University of Vienna to work with the eminent chemist, Ernest Späth. He received his PhD in 1931, having dazzled the Viennese with his skills at tennis and piano, fallen in love with opera, and acquiring a long-term collaborator, Josef Pikl.

Julian and Pikl returned to DePauw, taught chemistry, and within four years came up with the total synthesis of physostigmine from the calamar bean (*Physostigma venenosum*). Physostigmine was for many years the only weapon doctors had to fight glaucoma. The bean also contained stigmasterol, an intermediate in sex steroid synthesis, and Julian sought a more abundant source of plant sterols. He wrote to the Glidden Company (a natural product giant) requesting gallons of soybean oil. This contact led to a job interview at Glidden's labs in Appleton, Wisconsin. But Appleton had a hoary statute on its books dictating that "No Negro should be bedded or boarded in Appleton overnight." Chance favored the prepared chemist and Julian was offered a far better job in Chicago as Director of Research of the Soya Products Division of Glidden. The rest is chemical history. In more than eighteen years at Glidden, Julian developed "Lecithin Granules," Glidden's soya oil, Durkee's edible emulsifiers and—not incidentally—worked out the commercial syntheses of testosterone and progesterone from soybean oil. He used soy proteins to coat and size paper, to make cold water paints practical, and to size textiles. During World War II Julian invented AeroFoam, a soy protein product that quenches gasoline and oil fires; the foam saved lives from Europe to the Pacific. Julian was granted more than 100 chemical patents and Big Pharma still prepares hydrocortisone from compound S à la Julian, 1950.[7] In 1953 he founded his own company, Julian Laboratories, Inc., with labs in the US and Mexico. In 1961, the company was sold to Smith Kline & French for $2.3 million, "a staggering amount for a Black man at that time."[12]

In his lifetime, Julian was honored by membership in the National Academy of Sciences, a US Postal Service stamp, a dozen honorary degrees, directorships galore, and three public schools that bear his name. We also remember that this agile chemist made it possible to make cortisone from beans instead of bile so doctors could give it to patients for a pittance.

The Case of the Floppy-eared Rabbits

THE RABBITS, THREE OF THEM, are small and very white. They sit huddled in their cage, which in turn sits on a wooden trolley rolled up to a laboratory table. It is the spring of 1955 on the fifth floor of the Medical Science Building of NYU and light pours in from the East River. The new professor of pathology, Lewis Thomas (age forty-two), pulls one of the albinos out of the cage and places it on the black-topped lab bench. He turns to the group of second-year medical students perched on stools around the bench and asks them:

"Notice anything?"

They don't, immediately.

"It's a healthy bunny, if that's what you mean" one of students volunteers.

The professor smiles in reply: "You know, I didn't notice anything either when I first did this a few years ago. But last night, I gave this little fellow some papain by vein. Let's sit him next to one that hasn't been injected with papain. Here. Look."

He pulls out another rabbit. "Here's the control." Finally, he reaches for the third. "And here's another rabbit I also injected with papain."

The students look at the two who've been given papain side-by side with the control. Now comes the burst of recognition:

"Of course; Gosh!"

"The papain bunnies' ears are droopy!"

"I'll be darned, they look like dachshunds!"

"No, spaniels."

"Will they be alright?"

"What else happens? "

"What is papain?"

"What made you inject the animals with papain?"

"Is it cartilage that's wilting?"

Lewis Thomas does his best to answer the questions. He's done this many times before and it happens like clockwork. Sure, the rabbits will be fine. In a day or so the droopy ears will become erect again. Three days later, you won't be able to tell the papain bunnies from the controls. No. He's found no other ill effects of any kind in rabbits given papain. He has looked at sections of ears from the injected animals for days on end under the microscope and found nothing of interest. Nothing. Papain? It's an enzyme from the papaya plant that is used to break down proteins of all sorts. A protease. In the old days brewers used proteases like papain to clarify beer; nowadays papain is best known as a meat tenderizer. Can it break down cartilage? He doesn't know. But he guesses that cartilage is a quiet, inactive tissue. Perhaps only this dead sort of tissue responds to an injection of meat tenderizer. But, why inject rabbits with papain in the first place? Well, it's a long story, but it still is the most reproducible phenomenon he's ever seen in the lab! The students are vastly amused.

When Lewis Thomas recalled this episode a few years later he explained that he "couldn't really explain what the hell was going on."[1] But, trying to answer those questions got him thrashing around for explanations in the lab.

> I was in irons on my other experiments. I was not being brilliant on my other problems. . . . Well, this time I did what I didn't do before. I simultaneously cut sections of the ears of the rabbits after I'd given them papain and sections of normal ears. This is the part of the story I'm most ashamed of. It still makes me writhe to think of it. There was no damage to the tissue in the sense of a lesion. But what had taken place was a quantitative change in the matrix of cartilage. The only way you could make sense of this change was simultaneously to compare sections taken from the ears of rabbits which had been injected with papain with comparable sections from the ears of rabbits of the same age and size which had not received papain.[2]

Making sense of this took quite an effort. Two hundred and fifty rabbits were sacrificed, hundreds of sections were taken, but in the end, differences between comparable sections under the microscope were clear and striking. Although sections from the ears of papain-treated rabbits showed perfectly normal cells, the blue-staining material between cells seemed to have melted away. Ever since the days of Paul Ehrlich professional pathologists have called that material between cells the "basophilic matrix" because it latched on to blue, basic dyes. Papain had broken down that matrix, the extracellular ground substance that makes all cartilage semirigid. Take your own ear between your thumb and index finger and fold it into a flute. It will spring back by itself: that's because the matrix is intact.

Papain had no effect at all on the health of cartilage cells in the bunnies' ear and there was no evidence of inflammation since no new cells had leaked from injured blood vessels to attack the cartilage. Thomas correctly deduced that the dramatic ear-droop was due to a chemical attack by papain on the basophilic matrix of cartilage. On the morning after the rabbits had received papain, most of the matrix had been leached out. Happily, when the ears snapped back to normal in a few days, the blue-staining material was back in force. Thomas figured out that cartilage, far from being a dead or inert tissue, could survive a withering attack and recover its form. In fact, cartilage could make new matrix by the earful!

Thomas immediately understood the implications of this finding for human disease. By the 1950s it was known that in osteoarthritis (the arthritis of old age) loss of cartilage matrix preceded inflammation and joint destruction; in other words, cartilage broke itself down. In rheumatoid arthritis, on the other hand, cartilage matrix was attacked by inflammatory cells drawn from the bloodstream—much as in an infected joint.

Thomas reasoned that tissue injury in general was due to the uncontrolled release of the body's own papainlike proteases, our own meat tenderizers as it were, whether released from tissue itself or from white cells attracted from the circulation. He was ready to go to press. In 1956, the work appeared in the *Journal of Experimental Medicine* and its opening lines became an instant classic of scientific description:

> For reasons not relevant to the present discussion rabbits were injected intravenously with a solution of crude papain, and the following reactions occurred with unfailing regularity: Within 4 hours after injection, both ears were observed to be curled over at their tips. After 18 hours they had lost all of their normal rigidity and were collapsed limply at either side of the head, rather like the ears of spaniels. After 3 or 4 days, the ears became straightened and erect again. . . . Apart from the unusual cosmetic effect, the animals exhibited no evidences of systemic illness or discomfiture, and continued to move about after the fashion of normal animals of the species.[3]

Two additional observations were included in the original paper, one of limited interest, the other with major consequences in the clinic. Papain exists in both latent and active forms: products of living tissues can activate the latent enzyme. Thomas was surprised to find that cartilage was able to reactivate the enzyme: therefore cartilage is not a "quiet, inactive tissue." But, more importantly, Thomas discovered why cortisone has some very bad side effects. When he treated rabbits with cortisone after injecting them with papain, the animals' ears remained limp as long as they continued to receive cortisone. This was strong proof that cortisone inhibited repair of cartilage matrix. We now know that drugs that block repair of cartilage are bad news for bone and cartilage. Indeed, when cortisone inhibits the synthesis of matrix proteins of patients with arthritis, their tissues become fragile and their tender bones easy to break. Thomas's rabbits told us why.

But it wasn't the action of cortisone that drew public attention to the floppy-eared bunnies. It was that "unusual cosmetic effect" that caught the fancy of the public, thanks to pictures in the *New York Times* and the *New York Herald Tribune*. "An accidental sidelight of one research project had the startling effect of wilting the ears of rabbits" wrote Harold Schmeck in the *Times*.[4] The droopy-eared rabbits were featured in *LIFE* magazine, "Lop-eared Lapines" the journal tagged them,[5] and reporters flocked to NYU. The journalists wanted to know what the research project was of which these bunnies were the sidelight. Why was that research "not relevant to the present discussion?" What was papain? And why did Thomas inject animals with papain in the first place?

The papain story was a long story, indeed. It started in September of 1945 on the island of Guam where Thomas was serving with the Rockefeller Institute Naval Medical Research Unit. The war in the Pacific had ended in August and Thomas expected to be shipped home immediately, but three long months were to pass before the Navy shipped him stateside. As luck would have it, Thomas had no further official tasks, the Unit had unused research facilities and equipment, and there was an ample supply of laboratory animals. The bacteriology lab was fully supplied with media and stock cultures of many microbes. So Thomas went to work on a problem that had puzzled him since medical school, the pathogenesis of rheumatic fever. He spent the war years attempting, with varying degrees of success, to produce changes in hearts of experimental animals that looked like rheumatic fever in humans. And for a decade thereafter he tussled with the problem: in war and in peace, and as he moved up the academic food chain from Tulane, to Minnesota, and finally to New York.[6]

Thomas reasoned that known proteases, when injected into the blood stream of experimental animals, should induce lesions that mimicked those of rheumatic fever. As he recalled in 1958, he did the experiment because:

> It's an attractive idea on which there's little evidence. . . . For this investigation I used trypsin, because it was the most available enzyme around the laboratory and I got nothing. We also happened to have papain; I don't know where it had come from; but because it was there, I tried it. I also tried a third enzyme, ficin. It comes from figs. . . . So I had these three enzymes. The other two didn't produce lesions. Nor did papain. But what the papain did was always to produce these bizarre cosmetic changes [the floppy-eared rabbits]. It was one of the most uniform reactions I'd ever seen in biology. It always happened. And it looked as if something important must have happened to cause this reaction.[7]

However, shortly after Thomas arrived in New York he realized that he had come to a dead halt in a related line of research, a puzzle that took the wind out of his sails and prompted his confession that "I was in irons on my other experiments. I was not being brilliant on my other problems. . . ." "In irons" sounds about right, he was making no headway in figuring out how cortisone might work. Lewis Thomas, like other doctors at midcentury, was convinced that cortisone, a drug that ranked with insulin, penicillin, and vitamin B_{12} as one of the medical wonders of the age, was the long-sought answer to rheumatic diseases. It was and it wasn't.

And so, five years before he showed those floppy-eared rabbits to his students in New York, he set briskly to work on cortisone in Minnesota. By 1951 Thomas had published six papers on cortisone's effect on experimental tissue injury and inflammation, but, alas, the results were less than conclusive. He confessed that he had collected a "good many facts," but had gained little real understanding; the action of cortisone remained a puzzle. Whereas cortisone clearly stopped inflammation dead in its tracks in the clinic, experiments in the lab seemed to show just the opposite. Indeed, it even helped microbes to spread and made their toxins more lethal.[8] Why did the wonder drug provoke in the lab the very kind of inflammation it had been devised to combat in the clinic? As they stood, the results seemed improbable, if not impossible.

Although his meanderings with cortisone weren't all that laughable at the time, Thomas' "disquieting sense of being wrong" made him look about for something new. And while the first phrase in the title of his landmark paper was "Reversible collapse of rabbit ears after intravenous papain," he couldn't let cortisone go. Thomas had to see what would happen if he added cortisone to the equation. The full title was "Reversible collapse of rabbit ears after intravenous papain, and prevention of recovery by cortisone."[3] Thomas had expected neither that the ears would collapse nor that cortisone would prevent their recovery. When he rehearsed the full story of the rabbits with his coworker Bob McCluskey in 1958, Mac told him that the experiments illustrated Dwight D. Eisenhower's dictum that "all plans are useless, but planning is essential." I was a novice in the lab at the time, and was convinced that Thomas had simply pulled the rabbits out of a hat.

The papain experiment not only made headlines in the local papers, but coarse-grained photos of the droopy-eared rabbits were carried nationwide by the wire services. More immediately, the papain experiments made Thomas an instant celebrity at his new institution, attracting students and postdoctoral fellows from far and wide. At each retelling of the papain story, of explaining to his students and postdocs how he discovered that a plant protease caused rabbit ears to collapse, he called it a case of serendipity: "Serendipity is a familiar term. I first heard about it in Dr. Cannon's class at the Harvard Medical School."[9]

Eventually the rabbit-ear experiments became the subject of a scholarly article by two sociologists of science whose curiosity had been aroused by the *Times* article. For good reason they called their account "The case of the floppy-eared rabbits: An instance of serendipity gained and serendipity lost."[1] Bernard Barber and Renée Fox, of the University of Pennsylvania, interviewed not only Lewis Thomas, but also Aaron Kellner, a Cornell pathologist, who had injected rabbits with papain at about the same time as Thomas, and for many of the same reasons. Kellner was also working on rheumatic fever and had injected rabbits with a variety of proteolytic enzymes for several years in order to produce cardiac lesions that resemble rheumatic heart disease.[10] But Kellner had missed the boat. Like Thomas, Kellner had noticed that papain made rabbit ears droop; he told his interviewers that every technician he had ever employed knew that if the rabbit ears collapsed, the enzyme was bioactive. "We called them the floppy-eared rabbits," Kellner recalled, but "I didn't write it up." The sociologists explained why Kellner missed the obvious:

> However, for one of those trivial reasons that sometimes affect the course of research—the obviously amusing quality of floppiness in rabbits' ears—Dr. Kellner did not take the phenomenon as seriously as he took other aspects of the experimental situation involving the injection of papain.[11]

I'm sold on the notion that temperament is the mold of success, in science as in other endeavors, and in that respect Thomas was the most sanguine of us all. A keen appetite for the "obviously amusing" is an endearing aspect of that temper, which Thomas shared with his teacher, Walter B. Cannon. "Serendipity," Cannon had called it. "Serendipity gained and serendipity lost," wrote the sociologists. Another phrase might be just as apt. The title of a biography of Noël Coward, *A Talent to Amuse*,[12] comes to mind in Thomas's recollection of the floppy-eared rabbits:

> I was able to justify working on so seemingly frivolous a problem by the possibility that one might figure out how cortisone might work...which gave the whole affair a down-to-earth usable aspect. But, I was obliged to confess, despite this, that the work had been done because it was amusing.[13]

That's a pretty sound strategy for discovering the new. All it takes is a talent to amuse. And to remain amused.

Einstein and Jimmy Mac

TWO IMAGES STICK IN MY mind as icons of twentieth-century science; they illustrate two very different public lives and two very different ways of looking at science. A scant twenty years and eighty miles apart in origin, they link an Einstein letter to the atomic bomb and the atomic bomb to the human genome project.

The first image is a bust of Albert Einstein (age fifty-three), kneaded in clay by Sir Jacob Epstein in September of 1933. A score of bronze copies were cast, nowadays they preside over libraries on three continents. The physicist casts an avuncular smile at the onlooker; his face shows the rewards of intellect and *The Life It Brings*—to borrow a phrase from Jeremy Bernstein's warm memoir.[1] The bust recalls days when automobiles had running boards and when a limerick was current that now touches on the quaint:

> What a curious family called Stein
> There's Gert, there's Ep and there's Ein
> Gert's writing is punk,
> Ep's statues are junk
> And nobody understands Ein.

Gert (Gertrude Stein) and *Ein* (Albert) need no introduction, but the reputation of *Ep* has fallen so low that few now remember the works of Sir Jacob Epstein (1880–1959). The career of this American-born English sculptor began with hard-edged carvings in the vorticist vein and ended with high-priced portrait busts which look as if they had been cast from chopped liver.

Epstein modeled his bust from life at a cottage on the Norfolk coast to which Einstein had fled from Germany en route to Princeton's Institute for Advanced Studies. Then at the height of his own fame, Epstein recollected that "When I was doing Professor Albert Einstein's bust he had many a jibe at the Nazi professors, one hundred of whom had condemned his theory of relativity in a book. 'Were I wrong,' he said, 'one professor would have been enough.'"[2] That unlikely anecdote has darker undertones. Philipp Lenart, one of those probably less than one hundred Nazi professors had indeed condemned Einstein in May of 1933:

> The most important example of the dangerous influence of Jewish circles on the study of nature has been provided by Herr Einstein with his mathematically botched-up theories consisting of some ancient knowledge and a few arbitrary additions. . . . Even scientists who have otherwise done solid work cannot escape the reproach that they have allowed the relativity theory to get a foothold in Germany because they did not see, or did not want to see, how wrong it is, outside the field of science, also, to regard this Jew as a good German.[3]

Epstein's bust catches a wistful aspect of Einstein in those early days of exile, when the gates of German science were shutting. Einstein had resigned from the Prussian Academy of Science which had accused him of "agitational activities," his books had been burned before the *Staatsoper* in Berlin, and he never set foot in Germany again. And when Mussolini joined Hitler's racial crusade in 1938, Einstein resigned from Rome's Linceian Academy as well. In his Princeton years, he continued his quest for a theory that might unify relativity with quantum mechanics. He also became a guiding spirit of the émigré scientists who brought atomic physics to America, among them Leo Szilard, Enrico Fermi, Edward Teller, and Hans Bethe. Viewed by the larger public as a sage of science, he was an early advocate of civil rights and a staunch supporter of the Jewish state. Then, On August 2, 1939, he signed the first of four letters to President Franklin Roosevelt that started the clock ticking on the Manhattan Project:

> In the course of the last four months it has been made probable—through the work of Joliot in France as well as Fermi and Szilard in America—that it may become possible to set up a nuclear chain reaction in a large mass of uranium, by which vast amounts of power and large quantities of new radium-like elements would be generated. Now it appears almost certain that this could be achieved in the immediate future. This new phenomenon would also lead to the construction of bombs, and it is conceivable—though much less certain—that extremely powerful bombs of a new type may thus be constructed.[4]

Einstein's letter (largely drafted by Leo Szilard) went on to warn Roosevelt that the Germans were aware of these possibilities; they had already banned export of uranium from recently occupied Czechoslovakia. Roosevelt's prompt response to Einstein was to set up a three-man, scientific Advisory Committee on Uranium. The committee's efforts became more urgent after Pearl Harbor, and a crash program to develop the atomic bomb was placed under the Army Corps of Engineers. The Corps invented a new district, called the "Manhattan Engineer District" aka the Manhattan Project; major facilities were established at Los Alamos, Oak Ridge, and the University of Chicago. The physicists of the Manhattan Project, many of them émigrés themselves, were certain that German scientists were preparing to make those "extremely powerful bombs of a new type." The race was on against the Nazis.

Four decades later, scientists at the Los Alamos National Laboratory were again in a race: sequencing the human genome. In line with federal wisdom, they were now working for the Department of Energy, an agency mainly concerned with fossil fuel. The DOE had taken over nuclear matters from the short-lived Atomic Energy Commission (1946–1977), which had made the Los Alamos facility a world center of supercomputing.[5] In keeping with its cold war mission, the AEC had also collected vast amounts of data on the effects of atomic radiation on chromosomes, genes, and DNA. Into those computers went much of the data collected by the Atomic Bomb Casualty Commission (1946–1982). With the cold war ended in the 1980s, all that computer power and all those bytes of genetic data—not to speak of jobs in New Mexico—were about to become history. An administrator at the DOE, Charles DeLisi, realized that another Big Science project was in the offing. Prompted by farsighted proposals from biologists such as Robert Sinsheimer, Renato Dulbecco, and Sidney Brenner, DeLisi took the first practical steps toward the Human Genome Project. With support from Senator Pete Domenici of New Mexico, he diverted 5.3 million dollars of his 1987 budget for the preliminaries. Contentious scientific meetings in Santa Fe and at Cold Spring Harbor addressed the ways and means of the project; critics such as David Botstein of Stanford complained that this distracting effort at Big Science was "a program for unemployed bomb makers."[6]

Finally, on October 1, 1988, the Office for Human Genome Research was created within the National Institutes of Health. Both NIH and the DOE signed a memorandum of understanding to "coordinate research and technical activities related to the human genome." James Dewey Watson was appointed director of the first crash program in biology, which many at the time likened to the Manhattan Project. "To me it's crucial that we get the genome now rather than twenty years from now," Watson proclaimed."[6] The race was on.

The second of my photo icons is a photograph of James Watson and Francis Crick. Taken on May 12, 1953, by Antony Barrington Brown, a freelance photographer, it shows Crick and Watson posed by their new model of DNA in the old Cavendish labs at Cambridge; the Cavendish is just a short run down the A11 from Einstein's former cottage near Cromer.[7] On the last day of February 1953 Francis Crick had announced to the patrons of the nearby Eagle pub on Benet Street that "We have discovered the secret of life" and the double helix made its debut in *Nature* on April 25th. A photo session was arranged, for which the usually scruffy Watson had been prepped by a colleague who hoped to sell story and photos to *Time* magazine. Watson recalls that

> I was not trusted to act alone and Odile [Crick] accompanied me to the men's clothing shop across from the chapel of John's [the College]. My ill-fitting American tweed jacket was thrown out and replaced by a blue blazer and associated gray trousers. They would much better express my new status as the co-winner of a very great scientific jackpot.[8]

Four "snaps" were taken of the two with their model; *Time* declined to publish either the story or the pics and the iconic image found its way into *Varsity*, the local rag. On the occasion of the fiftieth anniversary of that discovery, a Cambridge blue plaque was placed by the Eagle's door and large blow-ups of the Brown photograph dominated celebrations around the world. In the original photographs a very young Watson (age twenty-five) looks almost goofy and dazed by what he and Crick have wrought. They'd won a race against the likes of Linus Pauling! Brown remembers that

> I was affably greeted by a couple of chaps lounging at a desk by the window, drinking coffee. "What's all this about?" I asked. With an airy wave of the hand one of them, Crick I think, said, "we've got this model" indicating an array of retort stands holding thin brass rods and alls. Although supposedly a chemist myself it meant absolutely nothing to me and fortunately they did not expose my ignorance by attempting to explain it in terms I might just have comprehended. Anyway, I had only come to get a picture so I set up my lights and camera and said, "you'd better stand by it and look portentous" which they lamentably failed to do. . . .[9]

Watson certainly looks lamentably less than portentous, he appears anxious, driven, and very much the sort of chap whose entrepreneurial accounts of science in *The Double Helix*[10] and *Genes, Girls, and Gamow*[8] were to gain him the jealous contempt of his peers. E. O. Wilson, the environmental biologist and Watson's colleague at Harvard in the 1960s, publicly described Watson as the most unpleasant human being he had ever met—the "Caligula of Science."[11] Contrast that with Abraham Pais's recollection that: "Einstein's company was comfortable and comforting to those who knew him."[12] One cannot imagine any scientist of Einstein's generation confessing, as Watson did to an interviewer, that "I really was brought up without the inhibitions of religion or good manners."[13] Nor could one imagine Einstein crowing, as Watson did to a colleague "I do not have *a* Nobel prize, I have *the* Nobel prize."[6]

And yet . . .

I've always marveled how dominant a role Einstein played in the public imagination: he bestrode the science of the twentieth century, as Newton had of the eighteenth. Was it the impact of relativity? I doubt it: the limerick is right, nobody understood Ein. Was it the equation $E=mc^2$ and that letter to Roosevelt? Perhaps. His paternity of the bomb certainly boosted his postwar public image, but Einstein was Einstein before the Manhattan Project. Indeed, after the war, Einstein regretted having signed Szilard's letter: "Had I known the German would not succeed in producing an atomic bomb, I would not have lifted a finger."[14] Was it his image in the tabloids and newsreels: a frizzy, white-haired, violin-playing, avuncular exile in a sweatshirt who assured one and all that God did not play dice with the universe? Possibly all of the above, but chiefly, I daresay, because he was better than anyone else in his century at not only disturbing, but also consoling, the universe—to paraphrase Freeman Dyson.[15] Einstein, whose name had become synonymous with genius, reassured the laity that "A legitimate conflict between science and religion cannot exist. . . . Science without religion is lame, religion without science is blind."[16]

Biology has had few such towering figures. Charles Darwin and Sigmund Freud, the Einsteins of genes and behavior, have changed the way many of us look at the world, but their work cannot be said to have gained the sort of universal assent won by quantum mechanics. Those who can read the equations—whether in Oak Ridge or Teheran, Pyongyang or Islamabad—probably agree on the consequences of $E=mc^2$; the same cannot be said of the theory of natural selection or a dynamic unconscious. Theology may some day lie down with molecular biology, but—as

Woody Allen said of the lamb and the lion—it's not likely to get a good night's sleep. On the other hand, cosmology offers the nightcap of consolation: compare Einstein's aphorism "*Raffiniert ist der Herrgott aber boshaft ist er nicht*" ("God may be subtle but he's not malicious")[17] with Watson's challenge:

> With its direct contradiction of religious accounts of creation, evolution represents science's most direct incursion into the religious domain and accordingly provokes the acute defensiveness that characterized creationism.[18]

I'm persuaded that in the unlikely figure of James Dewey Watson our new biology has thrown up a hero to vie with the cosmologists. But, unlike Einstein, he throws no sop to the Gods. The news he brings is tougher to bear than that of relativity, which in any case was no slap at our collective egos. It's not the news of a heliocentric universe; that was Galileo's message. Nor is Watson's story a dispatch from the *Beagle*; thanks to Darwin and classical genetics we already know from whence we came. No, Watson steps before the cameras to tell us that we're made of simple chemicals and that he's figured out how they're arranged. Watson's answer to the cross is the double helix of DNA:

> It brought the Enlightenment's revolution in materialistic thinking into the cell. . . . The double helix is an elegant structure, but its message is downright prosaic: life is simply a matter of chemistry.[18]

That's a hard message, and wise, dyskinetic James Watson is not the most genial of messengers. "Lucky Jim" he's been called, "Johnny Mac" he fancies himself. In a rare flash of self-analysis, Watson told an interviewer that in a film version of *The Double Helix*, "I should be played by John McEnroe. . . . You know, someone who pisses people off."[13] Not immodest, considering that Watson has won a number of Grand Slams in science: he's solved the structure of DNA, written a classic memoir of the solution, posed the riddle of RNA, mentored the Human Genome Project, and built a small Long Island laboratory into a world center of molecular science. Moreover, his writings—from the autobiographical *The Double Helix*, to the encyclopedic *DNA: The Secret of Life*—have also been at the Grand Slam level. Performances worthy of Johnny Mac, they've been deft displays of science writing by a top-seeded scientist. They seem almost purposefully written to piss off the politically correct, the fans of revealed religion and eco-sentimentalists of any stripe. Nor has Watson patience with whims of the high and mighty:

> Let me be utterly plain in stating my belief that it is nothing less than an absurdity to deprive ourselves of the benefits of GM foods by demonizing them; and, with the need for them so great in the developing world, it is nothing less than a crime to be governed by the irrational suppositions of Prince Charles and others.[19]

McEnroe, whose center-court antics before the Royal Box did not interfere with seven Grand Slam tournament victories, is a fit model indeed. The *New York Times* never compared Johnny Mac to Caligula, but was moved to call him "the worst advertisement for our system of values since Al Capone."[20] I find Watson's case for reductionist science in *DNA: The Secret of Life* the best advertisement for our system of Enlightenment values since Denis Diderot.

> When discussing our genes, we seem ready to commit what philosophers call the "naturalistic fallacy," assuming that the way nature intended it is best. By centrally heating our homes and taking antibiotics when we have an infection, we carefully steer clear of the fallacy in our daily lives, but mentions of genetic improvement have us rushing to run the "nature knows best" flag up the mast. For this reason, I think that the acceptance of genetic enhancement will most likely come about through efforts to prevent disease.[13]

I'd also agree with Nancy Hopkins, a professor of biology at MIT and a student of Watson's when he taught at Harvard during the sixties that "He's one of those few people who changed the world in a way people will be talking about 100, 200, 300 years from now."[21] And on the fiftieth anniversary of the discovery of the three-dimensional structure of DNA, Watson presented us with *DNA: The Secret of Life*, perhaps the best popular exposition of how we are put together as humans. It offers a blueprint of molecular genetics from sweet peas to stem cells; it reminds those who grew up with the biological revolution of its history; and, finally, it recapitulates the necessary argument for reductionist science:

> The discovery of DNA put an end to a debate as old as the human species: Does life have some magical, mystical essence, or is it, like any chemical reaction carried out in a science class, the product of normal physical and chemical processes? Is there something divine at the heart of a cell that brings it to life? The double helix answered that question with a definitive No.[13]

It's no wonder that the man who assembled the first explanatory text of modern biology, *The Molecular Biology of the Gene* (1965), has provided the rest of us with easily comprehensible arguments for value-free science. Watson's plea for the unbiased study of behavioral genetics, for example, is not irrelevant to self-censorship in scientific journals today:

> Let us not allow transient political considerations to set the scientific agenda. Yes, we may uncover truths that make us uneasy in the light of our present circumstances, but it is those circumstances, not nature's truth, to which policy makers ought to address themselves.[22]

But Watson's reductive approach is no reversion to pre-DNA genetics or simplified dog-eat-dog survival theories. He has looked hard and long at the nasty legacy of the eugenics movement. Eugenics was first professed in Anglo-Saxon lands by social Darwinists such as Charles Davenport, who—ironically—served as the first director of the Cold Spring Harbor Laboratory. The eugenics movement acquired eliminationist rhetoric as it flourished under the Fascist and Nazi regimes; it ended at Auschwitz. Meliorists and Fabians, on the other hand, promoted eugenics of a temperate kind; they hoped to cleanse the gene pool by tidy breeding based on tight statistics. Watson reminds us that Francis Galton, Darwin's cousin, disproved the notion that prayer could prevail over genes. Despite centuries of Sunday prayers for the monarchs of Britain, the cumulative effect of all those prayers was not beneficial: "On average the monarchs died somewhat younger than other members of the British aristocracy."[23] (see "Galton's Prayer)

Watson's recent accounts of DNA history are more evenhanded than in his jejune memoirs *The Double Helix* and *Genes, Girls, and Gamow.* He now inclines to Claude Bernard's distinction between art and science: "*l'art c'est* moi, *la science c'est* nous." Watson may himself be moving from *moi* to *nous*. In *DNA: The Secret of Life* he tells us that genetics became chemistry thanks to J. F. Miescher's isolation of nuclein from pus (1871) but that DNA didn't become "the secret of life" until Oswald Avery, Colin McCleod and Maclyn McCarty transformed the genes of bacteria with pure DNA in 1944. By 1953, with A-T, G-C base ratios established (Erwin Chargaff), with crystallographic patterns of dry and wet DNA available at a glance (Rosalind Franklin, Maurice Wilkins), and with enol-ketone tautomers to explain base pairing (Jerry Donohue), Watson and Crick did the thought experiment of the century and came up with the model shown in the Barrington Brown photograph. That plaque on the wall of the Eagle proclaiming "the discovery of the secret of life" might well list many other "we's" who made and extended the discovery of the double helix.

One could, indeed, argue that there are many "secrets of life." If life is just a matter of physics and chemistry, what about the physics and chemistry of how mitochondria trap energy from the sun using ATP (Peter Mitchell, John Walker), of how amphipathic lipids trapped organic molecules from the primal soup (Alec Bangham), or how $n\lambda=2d\sin\theta$ (the Bragg equation) permits crystallographers to decipher matter? Each of those secrets also popped up by the river Cam.

Watson's new story of DNA is illustrated by a collection of photos that should make the top ten of anyone's icon list. There is an impish snap of a young Matt Meselson at a Caltech ultracentrifuge when he and Frank Stahl carried out the "most beautiful experiment in biology": the semiconservative replication of DNA. Writes Watson, "They met in the summer of 1954 at the Marine Biological Laboratory at Woods Hole, Massachusetts, where I was then lecturing [in the Physiology Course], and agreed—over a good many gin martinis—that they should get together to do some science."[24] The many we's of science formed a web. Francis Crick and Rosalind Franklin were all there that summer. As the Lasker Foundation has noted

> It has been said that networking within the research community separates the vast majority of scientists by only a small number of people, perhaps only five or six (as in the play, *Six Degrees of Separation*). Nowhere is this more apparent than in the 1950's, when the great thinkers—and founders of the gene business—from the labs at Cavendish, Cold Spring Harbor, Woods Hole, The Institut Pasteur, and Caltech planned their summers together in Woods Hole or Cold Spring Harbor, remaining in close contact for the rest of the year.[25]

And sure enough, Watson's version of the DNA photo album extends from Jacob, Monod, and Lwoff at the Pasteur, to the adolescent antics of the RNA tie–club (Alex Rich, Sidney Brenner, Leslie Orgel, Francis Crick, George Gamow, *et al.*) at Cambridge, Woods Hole, and Cold Spring Harbor. The album is a kind of highlight tour of science in the twentieth century.

The twenty-first began with another highlight photograph, that of Bill Clinton with Craig Venter and Francis Collins at Center Court (Washington) in June of 2000 to announce the winner(s) of the human genome tournament.[26] In the photo Collins, a born-again Christian, shares honors with a grinning, entrepreneurial Craig Venter. Clinton looks radiant. Speech after speech: Clinton describes the human genome as "the language in which God created Man." No one on that grand occasion brings up the Einstein/Szilard letter to Roosevelt, the Manhattan Project, the dreams of Sinsheimer, Brenner, and Dulbecco, or those supercomputers at Los Alamos. A wider-angle lens would have shown a somewhat forlorn James Watson seated in the shadow of Francis Collins, who succeeded him as director of the public genome consortium. Watson had resigned from his NIH appointment in 1992 to "spend more time with his lab"—in the George Tenent sense. His ouster from the project was ostensibly over the issue of gene patenting (he was against it), but I'd guess that irreverent Jimmy Mac had made no friends on the pious Potomac:

> Life, we now know, is nothing but a vast array of coordinated chemical reactions. The 'secret' to that coordination is the breathtakingly complex set of instructions inscribed, again chemically, in our DNA.[27]

Watson is pleased that humans turn out to have far fewer genes than first supposed. True to his reductionist belief, he dissents from Stephen Jay Gould's notion that fewer genes imply a holistic superstructure: "emergent qualities." Spotting the camel's nose of intelligent design poking into the tent of genetics, Watson argues that it's easier to sort through the chemistry and physics of 35,000 than 100,000 genes. Indeed, Watson's plainspoken arguments against petty piety and "nature knows best" politics are a far cry from the gentle pantheism—and gentler manners—of an Albert Einstein. His true lineage is that of Jacques Loeb, another legendary instructor in the Woods Hole Physiology course. Loeb, who was the first to create life in a dish argued in 1912 that "life, *i.e.*, the sum of all life phenomena, can be unequivocally explained in physico-chemical terms."[28] That's also a secret of life, but one based on guesswork rather than DNA, in those days before Jimmy Mac won the tiebreaker.

Baumol's Curse

YEARS HAVE PASSED SINCE I first received a letter addressing me as "Dear Health Care Provider" rather than "Dear Doctor." Trained to be skeptical of noun adjectives in New York City public school, I was at first amused by the salutation. Sad to say, it soon became clear that George Orwell was right to warn us that the aim of Newspeak is to narrow the range of thought. When medicine, a learned profession based on experience and experiment, became narrowed to "health care," its practice became the "health care system," and mischief followed upon mischief. As doctors became providers, patients were reduced to consumers; what a narrowing of roles! In consequence, in this country, and in one generation, the venal norms of commerce have become the narrow rules of the clinic. And then came "managed care."

Doctor and patient alike should have woken up when the furies came up with "managed care," which can only be defined as the denial of medical treatment by a doctor to a patient so that someone else can save money. Managed care means that doctors who should have signed up to heal all of the people all of the time are being asked to heal some of the people only some of the time and to watch the bottom line all of the time. In good measure, we've abandoned the traditional way of tending to the needs of one sick human being at a time, a tradition honored for at least two centuries as the "doctor/patient relationship" or, in French, "*le couple médecin/malade.*"

"The care in this country is still second to none," writes a past president of the AMA.[1] He's probably right about American medicine as a technical achievement: these days we *are* second to none when it comes to transplanting hearts, replacing hips, scraping clots, stemming leukemia, and raising healthy babies from a dish. Americans don't shop overseas for medical miracles, but kings, sheiks, and moguls come here for their cure.

Doctors and postdocs from every corner of the earth flock to us for advanced training. We've stopped SARS, dengue fever, and cholera at our borders; we've added over a decade to our life span since midcentury.[2] We dazzle the world in molecular science, nanotechnology, and pharmacogenomics; we've walked away with the glittering prizes.

But that's only part of the story. Our complex, confusing health care delivery system prevents us from providing that "second to none" medical care to all who need it. Medicine, the old art and modern science of taking care of patients and preventing disease, costs money, and that cost is becoming prohibitive. We spend almost 15 percent of our gross domestic product on health care, much more than any other nation. Last year, our government alone spent $252 billion on Medicare and another $28 billion on health research and training (the latter more than all the nations on earth combined.)[3] By the end of the decade Medicare will be up to $456 billion. These costs are such that industry leaders from GE to GM urge us to "crack down on medical costs through the same quality-control techniques employed on the factory floor."[4] And crack down, we have.

Our health care system, fragmented into HMOs, PPOs, the Blues, Medicaid, Medicare A and B, C and D, the "for profit" and "not for profit" hospitals, copays and pay limits, etc., is already cunningly designed to ration medical care by techniques used on the factory floor. Car manufacturers control costs by increasing productivity (*i.e.*, shortening the man-hours required to make each car) but our health care system controls costs by limiting access (*i.e.*, shutting down the assembly line). Access to care is especially difficult for the uninsured. Those 60 or so million of our citizens who are not covered by insurance will, of course, receive some kind of medical attention, say, at Bellevue Hospital. But, too often, they seek care late in the course of diseases such as diabetes, hypertension, or eclampsia that have run wild for lack of prevention. Our system seems to combine the efficiency of socialism with the compassion of capitalism.

Our medical arrangements appear so messy and unfair, compared to the social democracies of Western Europe with their tidy, single-payer and/or safety-net systems in place for all. It's not surprising, therefore, that fully one-third of Americans believe that health care should be entirely reformed, compared to only 5 percent of Swedes and Danes. But it's not quite that simple, because Europeans stem the tide of medical costs by other means. In 1996, it took an average of 17 days for a patient with stable angina in New York state to receive a coronary artery bypass graft (CABG). In Sweden, patients with the same ominous pain in their chest had to wait 59 days, and Dutch patients waited 72 days! Nineteen of 3,500 Europeans died while waiting for their CABGs.[5] A recent Harvard study concluded that while American physicians worried that their patients couldn't afford appropriate medical care or essential prescription drugs, physicians in other democracies complained about waiting lists, a shortage of medical specialists, and inadequate facilities.[6]

The problem then is independent of any health care delivery system: in modern societies the cost of medical care is ordained to outstrip the resources available. Why? The answer is "Baumol's curse" or "Baumol's cost disease." Economist William Baumol is a distinguished colleague of mine at NYU, another Lynx, and a perennial candidate for the Nobel Prize. In 1966 (together with William Bowen, who later became president of Princeton) Baumol wrote a groundbreaking work on the economics of the performing arts, *Performing Arts: The Economic Dilemma.*[7] Baumol formulated the principle that in handicraft services such as the performing arts (and what he calls their "children," such as medicine or teaching), productivity has grown slowly as far back as there are records.[8] That means that as wages go up in any industrialized country, these handicraft services cannot increase their productivity enough to offset an inevitable rise in cost. Baumol adduced a now-famous example of the cost disease: in 1790 a Mozart string quartet (*e.g.*, the "Prussian" Quartet, K 590), required four musicians—two violins, one viola, one cello—and about thirty minutes of playing time to produce. The productivity of chamber music performance—output per person per hour—has not changed since 1790; but musicians now as then command a living wage. That's also true for medicine and teaching.

When I was in medical school in the 50s we were taught to take as much time as necessary to obtain a thorough history, perform a complete physical exam, and answer a patient's questions, and that's what I teach today. A seminar on Marcel Proust required one professor, fifteen students, and one hour weekly per semester at midcentury—and that's what it takes today. Consequently, the price of a concert ticket, a visit to a doctor, or a year in college has risen faster than the average good or service. Indeed, since 1950, doctors' fees have risen about double the rate of inflation, education expenditures per student nearly five times, and hospital costs seven times the rate. And doctors aren't getting any richer. Indeed, the real income of physicians has fallen slightly, slipping even relative to schoolteachers' pay.[9] Those of us in the handicraft industries can't increase our productivity fast enough; the curse is with us forever.

The measures devised to counter Baumol's curse are remarkably similar in single-payer countries and in hodgepodge America. Each operates on a provider/consumer model based on the faint hope that one can cut costs without hurting care. Many plans place a strict quota on the time that doctors can spend with their patients: in the Mozart analogy that's like playing the quartet in twenty minutes. In other plans, medical history–taking and patient advice have been handed over to paraprofessionals or to computer programs; that's like having an apprentice fiddler or a sound track accompany the Juilliard String Quartet. Most conglomerates will curtail lab tests and limit prescription drugs and specialist referrals; that's like forcing a musician to trade his Stradivarius for an eBay violin. But by and large, the most common practice is to force fewer doctors to see more patients; that's like dropping the cellist and asking the remaining three musicians to play a quartet.

Many date the onset of the provider/consumer model to 1962 when the covens of "health care planners" held their first Sabbath at the University of Michigan to announce the study of "health care delivery." On the eve of Baumol's curse, they declared that "The objective of all payors should be to obtain access to physician services for their subscribers at the lowest possible price. . . ."[10] But the curse has been operating longer than that. Troubled that American physicians might become servants of the state, as in National Socialist Germany, Hans Zinsser, the physician-writer, warned us in his 1940 autobiography, *As I Remember Him*:

> Now it is always relatively easy to conceive an ideal scheme or organization which shall represent the perfect mechanism for [medical care]. But such conceptions are likely to neglect certain imponderable values without which the machinery of service cannot run smoothly. In medicine, the problem is to find a solution which shall meet the requirements of effective scientific care of all those who require it, and retain at the same time that sense of personal responsibility, compassion and judgment without which the physician becomes a mere technician.[11]

What's surprising to me is that most doctors, in every system, in every industrialized country, manage to practice decent, conscientious medicine despite the constraints of Baumol's curse. A recent study from Columbia's School of Public Health[10] asked how doctors behave when some (but not all) of their patients are in managed care. It turns out that doctors spend approximately the same time with a patient whether or not that patient is under managed care or charged a fee for service. Doctors also prescribe pretty much the same medications, even if that means more administrative hassle in the case of those under managed care.[12] The harried doctors I know—in Boston and New York, Stockholm and Amsterdam—seem able to subvert the local system to make sure their patients get the best of care. When asked what makes them do it, to bend the rules, to spend the extra time, to outwit the gatekeepers on behalf of their patients, American physicians often respond, "I'm a doctor, I'm not a health care provider." They remain committed to the ethics of a profession rather than to the playbook of the factory floor.

When the black-bag doctors of my father's generation slogged their way through wintry house calls in the city they didn't expect to be second-guessed by bean counters at the other end of a telephone line. Nor did they expect teaching hospitals to hawk their wares in newspapers or on radio or television. And they didn't face ASK YOUR DOCTOR questions from folks made anxious by chemotherapy ads on the evening news. Their spirits were as willing as their drugs were weak. In the middle of the twentieth century doctors still plied their art much as they had in middle of the nineteenth, when Flaubert described the fictional Dr. Lariviere of *Madame Bovary* as one who

> cherishing his art with a fanatical love, exercised it with enthusiasm and wisdom . . . Disdainful of honors, of titles, and of academies, hospitable, generous, fatherly to the poor, and practicing virtue without believing in it, he would almost have passed for a saint if the keenness of his intellect had not caused him to be feared as a demon.[13]

Flaubert crafted this tribute as a sort of epitaph for his father, a distinguished practitioner of Rouen; it could have been written for mine, who died on a Manhattan street returning from a house call with snowflakes on his black bag.

That's why I'm reassured that in our own age of trade and hype, of HMOs and managed care, it remains possible, if difficult, for doctors to be "hospitable, generous, fatherly to the poor." In today's twilight of the one-on-one, in the sunset of the solo practitioner, there is still time for a saintly demon to practice virtue, with or without believing in it. Whatever the "health care system" in which they practice, it's good to know that most doctors have *not* become mere technicians, mere "health care providers." And the next time I hear Mozart's K 590, I'm confident that the quartet will still require four musicians—two violins, one viola, one cello—and thirty minutes to play.

From the Patchwork Mouse to Patchwork Data

> Summerlin pulled two white mice from the container. While they wriggled and squeaked in protest, he inspected the sites of the black skin grafts. Impulsively, Summerlin took his felt-tipped pen out of the breast pocket of his white coat and applied it briefly to the grafted patches on the two white animals. The ink made them look darker. Then he replaced the mice in the bin and strode out. . . .
>
> —From *The Patchwork Mouse* an account of William T. Summerlin's 1974 false claim of skin transplantation without immunosuppression.[1]

FRAUD WAS SO MUCH SIMPLER a generation ago. All one had to do was to take a felt-tipped pen and color a square patch of mouse skin. The incriminating patchwork was also easy to detect. Summerlin's faked "transplants" were discovered by a laboratory assistant who washed off the black ink with a ball of cotton soaked in a little alcohol. Yet the scandal and its upshot in 1974 were just as great as those aroused by the more complex frauds of today. The patchwork incident was described by Jane Brody in the *New York Times* as "a medical Watergate" that reflected "dangerous trends in current efforts to gain scientific acclaim and funds for research. Indeed, Robert A. Good, Summerlin's chief and coauthor, was accused of "manipulating national attention and attracting an enormous amount of money for the institute."[2] Soon afterward, Good stepped down as director of Sloan-Kettering, and that was the end of transplantation without immunosuppression. It sank without a trace in the literature; Summerlin's fraudulent papers have been cited only three times in the last twenty-five years; each a refutation.[3, 4]

Nowadays it takes more than a lab assistant with a cotton swab to detect fraud. In 2005 Seoul National University appointed a broad committee to investigate Professor Hwang Woo-Suk's claim that he had established eleven human embryonic stem cell lines by transfer of somatic cell nuclei.[5] The team reported that "The data in the 2005 article including test results from DNA fingerprinting, photographs of teratoma, embryoid bodies, MHC-HLA isotype matches and karyotyping have all been fabricated . . . the research team of Professor Hwang does not possess patient-specific stem cell lines or any scientific bases for claiming having created one."[6] To reach this conclusion, the committee collected cytogenetic samples, checked hundreds of DNA fingerprint samples, examined mitochondrial DNAs and compared dozens of polymorphic loci in the genes of alleged donors. Like Summerlin's patchwork fraud, Dr. Hwang's fabrication was attributed to scientific pride and overreach. Richard Doerflinger, of the United States Conference of Catholic Bishops, told the *New York Times* that "Hype and ambition have gotten ahead of the science."[7] Soon afterward, Dr. Hwang stepped down from directorship of his unit and that was the end of patient-specific stem cell lines in Seoul.

Remedies for Fraud

The worldwide response to Dr. Hwang's fraud was as predictable as that to the patchwork mouse a generation ago: pious hand-wringing and angry finger-pointing. Now, as then, critics of science blamed the messenger (the journal) rather than the message (the fabrication). Bold changes were demanded in the way science gets written, reviewed, and published, and scientific editors scrambled to get out of the line of fire. But one might say that science journals already have a Sarbanes-Oxley code in place. Most journals already insist that all coauthors sign off on the final manuscript, some oblige each author to spell out his exact role, others require each author to affirm the accuracy of every stage of the manuscript. Nevertheless, the editor of *Cell* suggests that it is time for the National Academy of Sciences to set strict new standards that apply to all.[8] My colleague, Donald Kennedy of *Science,* observed that: "A journal cannot go into authors' laboratories in search of fraud," but hedged his bet: "More actively, we are committed to examining our processes and ourselves in an effort to extract lessons for the future."[9] The *Journal of Cell Biology*, which already has an elaborate digital process in operation to detect image fraud, now plans to erect a screen of algorithms designed to spot specific types of image manipulation. The software was developed under a grant from the FBI.[9]

To determine whether the journal for which I am responsible, *The FASEB Journal,* should turn to image scanning, algorithms for detecting statistical swindles, etc., I've looked at recent cases of fraud and come up with some tentative conclusions:

1. Some scientists will cheat and we'll probably never know how often this happens. Young or old; MD or PhD; average or distinguished; male or female; black or white or khaki; some scientists will try to pull wool over the eyes of their colleagues, their reviewers, and their editors.
2. The culture of science is based on trust, not suspicion. Great scientists have had fraud committed under their noses, and good editors have published fabrication. Reviewers and editors must have a keen nose for swindle, but cannot engage in criminal investigation. As in political life, where I tend to side with liberty over security, in science, I'd go for trust over suspicion every time.
3. We ought not to rely on machines or algorithms of image or text analysis, but rely on the judgment of our editors, our editorial boards, and our reviewers as to whether a manuscript looks as if the data has been cooked. No measure of "quality assurance," no affidavit of authorship, no oath of responsibility or percent effort, can stop a soul hell-bent on self-destruction.
4. Fraud in science, if not in politics, is always self-destructive. Since the name of the game is confirmation, science is self-cleansing: flawed work is soon forgotten and remains uncited.

Those conclusions are based in part on three recent illustrations of the swindler's art.

Cancer of the Mouth

A study published by the Norwegian oncologist, Jon Sudbø, in *The Lancet* of October 15, 2005, concluded that long-term use of non-steroidal anti-inflammatory drugs like ibuprofen or naproxen could reduce the risk of oral cancer while exposing patients to higher risks of death from heart disease.[10] Sudbø's fraud was exposed when Camilla Stoltenberg of the Norwegian Institute of Public Health, who had access to the primary data, discovered that of the 908 people in the study, 250 shared the same birthday! After Sudbø confessed to his action, an embarrassed spokesman for the hospital admitted that the data were "totally false, actually totally fabricated. His database had been completely fabricated on his computer."[11] The fraud led editors at the *New England Journal of Medicine* to look into two earlier papers by Sudbø and sure enough, the *Journal* issued an "expression of concern" because two photomicrographs in a 2001 paper that purported to represent two different patients and stages of disease were in fact different magnifications of the same photomicrograph. Sudbø had patched bar graphs to overlay his photomicrographs so that the ruse was not immediately evident.[12, 13] Keeping with *The Lancet*'s antifraud policy, Sudbø's paper not only affirmed that "JS, JJL, SML, and AJD [Sudbø and collaborators] contributed equally to this paper" but also detailed the part each author played in statistic analysis, cutting the sections, writing the paper, etc. Ironically, as required, the paper affirmed that "All authors approved the final report." Conclusion: So much for quality assurance forms. And as to image fraud, Michael Rossner of the *Journal of Cell Biology*, while arguing that "The goal of a journal editor should be to catch these things before publication if at all possible," admitted that his methods are not yet up to catching tricks of magnification like Sudbø's.[13]

Immunologic Tolerance

Luk van Parijs, a thirty-five-year-old "rising star" in the field of RNA interference (RNAi), was dismissed from MIT in November of 2005. After the whistle was blown by coworkers in his lab, Van Parijs confessed that he had fabricated data in grant applications, published papers, and in submitted manuscripts. He'd been hired by MIT on the basis of strong publications as a graduate student at the Harvard Medical School and a productive postdoc stint with David Baltimore at Caltech. Now much of his work is under scrutiny. David Baltimore, Caltech president, and no stranger to such matters, wired *Science* magazine: "I thought Luk was an excellent scientist and truly cannot understand why he would fake anything."[14] *The New Scientist* engaged experts to scrutinize two of van Parijs's papers, one from Harvard, the other from Caltech.[15] The issue remains contested by van Parijs.[16] A 1998 paper in *Immunity* describes how *Fas/FasL* and *Bcl-2,* humoral factors relevant to cell death (apoptosis), affect our responses to self- and non-self recognition. *The New Scientist* review concluded that "Figure 1 of the paper contains 8 graphs. . . . Three graphs in the top row of the figure look very similar. Yet they are captioned as if they show data from three different mice. . . . In the bottom row, three graphs again look very similar. Yet they are again captioned as if they show data from three different mice." Worse yet was found in the Caltech paper:[17] "Two graphs in Figure 1C look very similar, especially if the graphs are printed out on transparent paper and superimposed. Yet one is captioned as if it comes from mouse cells infected with a human mutant gene, while the other is captioned as if it comes from mouse cells infected with a normal "wild type" human protein."[15] Conclusion: Even the best of senior scientists at Harvard and Caltech can have the wool pulled over their eyes, repeatedly.

Patching genes

In August of 1996, a reviewer for the British journal *Oncogene* found something strange on a submitted "Western blot" (a spot that shows a defined protein). The image accompanied a paper on cell transformation caused by an abnormal, fusion protein coded by an abnormal chromosome associated with acute myeloid leukemia. Coauthors of the paper included the recently appointed director of the US National Center for Human Genome Research (see "Reducing the Genome"), Francis Collins, and a young MD/PhD candidate at Michigan, Amitav Hajra. When the anomaly was called to Collins's attention, he immediately confronted the culprit. It was, Collins told *Science* magazine "one of those days you'll never forget."[18] Presented with evidence of a file of fraudulent data that Collins had unearthed, the student confessed. As the US Office of Research Integrity reported, Hajra fabricated data in "five published research papers, two published review articles, in one submitted but unpublished paper, in his doctoral dissertation, and in a submission to the national GenBank data base."[19] Collins circulated an apology to his colleagues in the field, withdrew the offending publications, and the government eventually exerted sanctions against Hajra. The inquiry had started with that patched Western blot. The reviewer for *Oncogene* noted that two lanes in the Western blot, labeled as two different proteins had the same background artifacts. Collins explained to *Science*: "There is a lane in the upper left, which if you cut it about halfway down, and then took the lower half and turned it 180 degrees so that the bottom is now the top, you end up with something that looks like a lane in the lower right." Not easy to spot, but "absolutely unequivocal once you look," Collins explained. Conclusion: It's hard to see the mote in your own eye, and it takes a sharp reviewer to cast it out.

I checked out the citation record for one of the retracted Hajra/Collins papers. As published in *Molecular and Cellular Biology* (vol. 15, pages 4980–9, 1995), the experiments would have constituted a dramatic test of the cancer-causing potential of the fusion protein. According to ISI, it's been cited only 12 times since 1995.[20] As a random control, I looked up the citation record of the publication that immediately followed the Hajra/Collins opus in the same issue of the journal (pages 4990–4997). It's a paper by a group of molecular biologists at McGill and also describes a protein/protein interaction that relates to nucleic acids and cancer.[21] That paper has been cited 276 times.

Conclusion

Science is self-cleansing, but check the Western blots.

Alice James and Rheumatic Gout

> . . . a lump I have had in one of my breasts for three months is a tumor. . . . This with a delicate embroidery of "the most distressing case of nervous hyperasthenia," added to spinal neurosis that has taken me off my legs for seven years, with attacks of rheumatic gout in my stomach for the last twenty, ought to satisfy the most inflated pathologic vanity.[1]

IN NOVEMBER OF 1882, after one of many "nervous" attacks, Alice James (1850–1892) took to her bedroom in the Boston home of her father, Henry James, Sr. On the advice of her brother, Dr. William James, and after ten days of bed-ridden solitude, she agreed to consult one of her brother's colleagues at the Harvard Medical School, Dr. Henry Harris Aubrey Beach. Dr. Beach suspected that there was "something lying in back of her nervousness," which coincided with her father's terminal illness. After three weeks of investigations, he informed his patient that her malady was "gout, rheumatic gout!"[2]

The diagnosis was relatively new on this side of the Atlantic, having been put on the medical map by A. B. Garrod of London in 1859. Beach, however, was a very up-to-date physician; he had taught anatomy under Dr. Oliver Wendell Holmes, held an appointment at the Massachusetts General Hospital, and was an assistant editor of the *Boston Medical and Surgical Journal* (now the *New England Journal of Medicine*). Beach promised relief to his patient and, indeed, she was relieved that her symptoms were due to something as physical as gout and not simply one more "fight between my body and my will."[3] But in December her father died; she again relapsed into general debility, muscle weakness, and depression. Eventually she placed herself in the Adams Nervine Asylum where for several months she was treated with electrical stimulators, vapors, and the rest cure of Dr. S. Weir Mitchell. These measures were of little avail and, after a variety of other treatments for symptoms attributed to "spinal neurosis," "nervous hyperaesthesia," "neurasthenia," or "suppressed gout," she crossed the ocean to join Brother Henry in England.

A point of irony is that in London she encountered one of the few physicians who could explain to Alice that rheumatic gout had little to do with her lifelong infirmity. Dr. Beach had arranged the consultation, assuring the Jameses that Sir Alfred Baring Garrod (1819–1907), was the only man who knew anything about "suppressed gout." Beach was somewhat off the mark. While Garrod was unquestionably the British authority on rheumatic complaints, he thought that the term "suppressed gout" was gibberish. Nowadays, Garrod is credited with turning the study of arthritis into the protoscience of rheumatology. At a public lecture on February 8, 1848, Garrod demonstrated that gout was due to a pile-up of uric acid in the blood and urine of the gouty, whereas there was no such increase in acute rheumatism (rheumatic fever).[4] At the time he was assistant physician at University College Hospital. Later, in 1854, he observed that linen strings dipped into the blood or urine of gouty patients became coated with visible crystals of uric acid. Garrod's "string sign" was a milestone in clinical biochemistry, similar deposits of crystals in the joints are the direct cause of gouty inflammation.[5, 6] Garrod went on to distinguish gout from rheumatic fever and, by the time he saw Alice James, he had just plucked a new diagnostic entity, rheumatoid arthritis, from the vague grab bag of conditions called "rheumatic gout."

> Although unwilling to add to the number of names, I cannot help expressing a desire that one may be found for the disease under consideration, not implying any necessary relation to gout or rheumatism. . . . I propose the term rheumatoid arthritis, by which name I wish to imply an inflammatory affection of the joints, not unlike rheumatism in some of its characters, but differing materially from it in its pathology.[7]

Alice James reported that she had "spent the most affable hour of my life" with Garrod, who told her in 1885 that the weakness in her legs and her digestive complaints were functional and not "organic" in origin. But —as usual with her doctors—she soon became disenchanted. At the dawn of the Freudian age she complained:

> I could get nothing out of him & he slipped thro' my cramped & clinging grasp as skillfully as if his physical conformation had been that of an eel instead of a Dutch cheese—The gout he looks upon as a small part of my trouble, "it being complicated with an excessive nervous sensibility" but I

> could get no suggestions of any sort as to climate, baths, or diet from him. The truth was he was entirely puzzled about me and had not the manliness to say so.[8]

Whether like an eel or a Dutch cheese in manly conformation, Garrod was sensible enough not to suggest changes in "climate, baths or diet." His patient suffered from no sort of gout known to him. Garrod was one of the first to draw clear distinctions between the common, heritable, and lead-induced varieties of gout. The lead-induced form, saturnine gout, had been rampant in the ruling classes of Europe since the sixteenth century and is caused by consumption of lead-contaminated spirits such as fortified wine or brandy (see "Galileo's Gout"). The disease, which in England reached epidemic proportions in the eighteenth century, took its greatest toll among the middle and upper classes, for the poor drank mainly gin and beer. Its epidemiology accounted for the notion of "the Honor of the Gout."[9] Since saturnine gout was due to Nurture, not Nature, Garrod and his contemporaries had a good notion of what to do for other forms of rheumatic complaints such as "rheumatic gout": change Nurture. Their well-off patients were packed off to take the cure at Bath, Leamington, or the continent, where a few weeks of lead-free water were expected to flush lead and uric acid out of their systems. Garrod so often prescribed hydrotherapy in Aix-les-Bains that today a street in the spa is named "Rue Sir Alfred Garrod."[10] Garrod had seen enough patients like Alice James to conjecture that in the potpourri of conditions called rheumatic gout "there is much to show in its etiology and the distribution of the affected joints that is intimately connected with the nervous system."[11]

Garrod and Alice James had a familiar sort of doctor/patient tussle with her symptoms. He believed that her muscle aches and cramps were caused by the galvanic stimulator she had used for several years; she believed that the Indian hemp (cannabis) prescribed by Garrod had made her sick. She sought refuge with a Dr. Townsend who informed her that she had a "gouty diathesis complicated by an abnormally sensitive nervous organization."[12] Finally, accompanied by her companion and a nurse, she took herself off to Royal Leamington Spa in 1889. She prepared to settle into the long, sessile routine of the invalid: *Alice in Bed* as legend has it these days.[13] Clinging to the belief that while some of her symptoms might be those of rheumatic gout, she blamed the rest on her own nervous temperament. "How well one had to be—to be ill!" she wrote in July of 1890.[14]

Nine months later, cancer struck; by the spring of 1892, she was terminally ill:

> I am being ground slowly on the grim grindstone of physical pain and on two nights I had almost asked for K's lethal dose, but one steps hesitatingly along unaccustomed ways and endures from second to second.[15]

The lethal dose would have been laudanum, and the "K" who dispensed it would have been Alice James's faithful companion, Katherine Peabody Loring. A few days after that entry Alice James was dead from metastatic cancer, killed by what she called "this unholy granite substance in my breast."

She had known full well that this was something palpably different from the afflictions that brought her to Leamington:

> To him who waits, all things come! My aspirations have been eccentric, but I cannot complain now that they have not been brilliantly fulfilled. Ever since I have been ill, I have longed and longed for some palpable disease, no matter how conventionally dreadful a label it might have. . . . It is entirely indecent to catalogue one's self in this way, but I put it down in a scientific spirit to show that though I have no productive worth, I have a certain value as an indestructible quantity.[16]

Indestructible she was not, but she continued nevertheless to produce in her journal an indestructible work of literature that ranks with those of her brothers. Discretion dictated that it appear eighteen years after Henry James's death (1934)[1] after passing in the family to the daughter of Robertson, youngest and frailest of the James brothers. The journal is a record of daily events, an almanac that condenses a decade, and a sharp text of self-knowledge. It had begun simply enough:

> I think that if I get into the habit of writing a bit about what happens, or rather doesn't happen, I may lose a little of the sense of loneliness and despair which abides by me.[17]

But soon she found her own style, in which social insight, cutting wit, and a good nose for literature mixed well with a bent for reform and a tendency to what brother Henry called "passionate radicalism." She was a sharp critic of conservative politics; a fan of Irish Home Rule and of Parnell. Hypocrisy outraged her:

> What a spectacle, the Anglo-Saxon races addressing remonstrances to the Czar against expelling the Jews from Russia, at the very moment when their own governments are making laws to forbid their immigration.[18]

She was also ahead of her time in sniffing out the cost of Empire. She worried about an English class structure in which:

> . . . the working man allows himself to be patted and legislated out of all independence; thus the profound ineradicables in the bone and the sinew conviction that outlying regions are their preserves, that they alone of human races massacre savages out of pure virtue. It would ill-become an American to reflect upon the treatment of aboriginal races; but I never heard it suggested that our hideous dealings with the Indians was brotherly love under the guise of pure cussedness.[19]

After her tumor was diagnosed, the tone of her journal turned somewhat darker and the entries more confessional. She permitted herself at last to express openly her love for Katherine Loring. Loring was a competent, educated woman of Brahmin stock, of whom Alice James had written to Sara Darwin (Charles Darwin's American daughter-in-law):

> She has all the mere brute superiority which distinguishes man from woman combined with all the distinctively feminine virtues. There is nothing she cannot do from hewing wood and drawing water to driving run-away horses & educating all the women in North America.[20]

Pictures show Katherine Loring as a tidy, angular woman: Alice B. Toklas as nurse/confidante. Loring had given up an active life to be with Alice James, shuttling between her own consumptive sister and the intermittently paralyzed James. In her last few months of life, the journal was no longer written by Alice James herself, but dictated to K. Alice James's tribute at what she called "this mortuary moment" is therefore perhaps more poignant:

> . . . is it not wonderful that this unholy granite substance in my breast should be the soil propitious for the perfect flowering of Katherine's unexampled genius for friendship and devotion. The story of her watchfulness, patience, and untiring resource, cannot be told by my feeble pen, but the pain and discomfort seem a feeble price to pay for all the happiness and peace with which she fills my days.

Reading Alice's journal of disease and despair, a tableau comes to mind. It is set in a Kensington drawing room that might have been painted by John Singer Sargent, but the feeling is pure Edvard Munch. It is January 6, 1892, two months before her death and Alice is racked by the jaundice of liver metastases. A photograph of the time shows Alice in her daybed; she looks drawn and gaunt, but has been dressed up and beribboned for the photographer. She sits stiffly propped on her cushions and appears pain-free for a while on morphine: the poppy is singing its song. She can no longer concentrate enough to write and dictates her journal entries to the beloved K. But K is by no means the only reader for whom these *pensées* are intended; she is also leaving a record for her brothers William and Henry.

Alice James contrasts the warm devotion of Katherine with the clinical cool of her doctors. Sir Andrew Clarke, who diagnosed that granite lump, is reprimanded for never quite being on time: he is of course the *late* Sir Andrew. But the eminent cancer specialist cannot help her, she is afraid that he is gripped by "impotent paralysis" and "talking by the hour without *saying* anything, while the longing, pallid victim stretches out a sickly tendril, hoping for some excrescence, a human wart, to catch on to . . ."[22] Alice James utters a cry from what John Keats called the World of Pains and Troubles to the world of her loftier brothers:

> When will men pass from the illusion of the intellectual, limited to sapless reason, and bow to the intelligent, juicy with the succulent science of life?[23]

Some day the science of life, biology, should be able to explain not only how tumor suppressor genes and hormones influence breast cancer, but also the lifelong affliction of souls as fragile as Alice James. The most convenient biological explanation is the genetic and keepers of the Jamesian flame point to a striking cluster of nervous pathology in two generations of the family. Deep depression, hysterical paralyses, and sexual ambiguity seemed to run in the clan: father Henry suffered from a Swedenborgian "midnight vastation"; Henry, Jr., was accused by William of "coddled sensibilities" and "dorsal anguish," William himself was paralyzed by "quivering fear." Mental homes provided refuge for the "incommunicable sadness" of youngest brother, Rob, and the "madness" of cousin Kitty.[24]

William James, the loftiest of the brood, believed that the family literally carried a genetic load on its back. Using the spinal imagery of his day, he wrote Rob: "I account it as a true crime against humanity for any one to run the probable risk of generating unhealthy offspring. For myself I have long since fully determined never to marry with anyone . . . for this dorsal trouble is evidently s'thing in the blood."[25] Alice also reverted to the serpent motif:

> Dr. Tuckey [a mesmeric doctor, recommended by William] asked me the other day whether I had written for the press, I vehemently denied the impu-

> tation. How sad it is that the purely innocuous should always be supposed to have the trail of the family serpent upon them.[25]

The family serpent is, of course, the double helix of depression and inspiration that ran through the James family, that "dorsal trouble . . . in the blood." The dying Alice James dictated to K an account of a visit paid her years before by Charles Darwin's daughter, Henrietta Litchfield. When Alice James told Mrs. Litchfield that her invalidism had for years been called "latent gout," the Darwin lady exclaimed "Oh! that's what we have [in our family] does it come from drink in your parents?"[26] The Darwins, like the Jameses, were a clan of morose geniuses, scholars, and medics who seem also to have had a family serpent in *their* blood. The Darwin biographies[27, 28] document a strong family history of depression, somatization reactions, and wives who took to their beds for decades; sure enough, much of this pathology was blamed on a "gouty diathesis." Depression and inspiration were closely intertwined in the Darwin line, as closely—one might say—as the two dancing serpents on the crest of Darwin College, Cambridge.

While social explanations rival the genetic, biography can describe, but not explain, the flight into sickness. Jean Strouse's authoritative biography[2] was written before paired lives like those of Alice James and Katharine Loring were explored in the context of legal constraints and homophobia. Treating that subject by elision, Strouse argued plausibly that Alice James's flight into disease was at least in part a not uncommon female strategy for coping with oppressive, male society in general and her father's expectations specifically. Strouse attributes some of Alice's symptoms to a daughter's reaction to a "kind father who had so blithely stimulated and thwarted her." She quotes Alice James's recollection of an acute early episode:

> As I used to sit immovable reading in the library with waves of violent inclination suddenly invading my muscles taking some one of their myriad forms such as throwing myself out of the window, or knocking off the head of the benignant pater as he sat with his silver locks, writing at his table, it used to seem to me that the only difference between me and the insane was that I had not only all the horrors and suffering of insanity but the duties of doctor, nurse and strait-jacket imposed upon me, too.[29]

In retrospect, no one can blame with confidence either Nature or Nurture for Alice James's life of infirmity, her unasked-for crucifixion, to use a phrase of Oliver Sacks's. Nurture in the form of nineteenth-century medical practice was clearly responsible for the diagnoses pinned on Alice James in her lifetime: spinal irritation, neurasthenia, hysteria, suppressed gout, etc. Medical practice was also responsible for subjecting her to electric prods, Indian hemp, spinal manipulation and all those buckets of tepid spa water . . . "as if to be singed and scalded were a costly privilege and leeches were a luxury" a phrase of Dr. Oliver Wendell Holmes.

Alice was one of many poor spirits who were flogged from pillar to post on the premise that their paralyses and palpitations were due to rebel humors in the spine; the tale of their prodding, buzzing, and poking has been told by the historian of our psychosomatic era, Edward Shorter.[30] Those treatments were not simply an assault of male doctors on their spinally challenged female patients. The male version of hysteria was called hypochondriasis and the paralyses it produced were also, willy-nilly, attributed to spinal irritation. Fresh from the Harvard Medical School, Dr. William James advised the most nervous of the James brothers, Rob, to take iron, to exercise, and to apply enough iodine to his back until his skin peeled.[24] The doctrine of "counter-irritation" demanded the show of wounded flesh on the part of male and female patient alike. That doctrine remains a major feature of folk medicine the world over; Western medicine abandoned it when we gave up "cupping."

In the end, her mother Mary knew what was wrong with Alice James: "It is a case of genuine hysteria for which no cause as yet can be discovered."[31] Alice herself awaited a new science of the mind. In her last letter to William, whose *Principles of Psychology* had just been published, she pleaded ". . . so when I'm gone, pray don't think of me simply as a creature who might have been something else, had neurotic science been born."[32]

These days, we may be better at the diagnosis and treatment of breast cancer, and the laws of Massachusetts have changed with respect to paired lives, but we are not much further along in neurotic science. From patients humbler than Alice James, we've learned that one does not need a "palpable disease" to be in real pain. Patients with psychosomatic diseases are often in anguish, and good doctors address the anguish and not the diagnosis. My own guess is that nowadays Alice James's illness would have been diagnosed as fibromyalgia, chronic fatigue syndrome, irritable bowel syndrome, or another one of man's "medically unexplained diseases." The British psychiatrist, Simon Wessely of King's College, London, explains that most medical specialities define unexplained syndromes in the technical terms of their own specialty. Presented with the same cluster of symptoms by a patient, what a rheumatologist will call "fibromyalgia," a gastroenterologist would diagnose as "irritable bowel syndrome," while a neurologist might come up with "chronic fatigue syndrome," and an infectious disease specialist might test for "chronic Lyme disease." Wessely provides convincing evidence that none of these monickers describes a unique clinical entity; each syndrome shares much with all the others. From epidemiological evidence he concludes that there are strong associations between persistent symptoms such as muscle weakness, stomach cramps and overall fatigue—the symptoms of Alice's "rheumatic gout" one might say. A safer bet, he argues is to describe these conditions honestly as "medically unexplained syndromes."[33, 34]

Unexplained or not, one more set of medical diagnoses intrudes on the story. A. B. Garrod left us a legacy as imposing as his work in the rheumatic diseases. His son, Sir Archibald Edward Garrod (1857–1936) followed his father's teachings on the heritable nature of gout. He became interested in a series of rare genetic disorders that, in his Croonian Lectures, he called "Inborn Errors of Metabolism."[35] And it is thanks to modern studies of inborn errors of metabolism—phenylketonuria is a prime example—that the principle emerged of "one gene, one enzyme." By means of that principle, Arthur Pardee, François Jacob, and Jacques Monod were able in 1959[36] to discover how changes in nutrients outside a bacterium can switch single, transmissible genes on or off within a cell; an example of how Nurture modifies Nature in the dish. It was also one of the milestones of molecular biology.

Nowadays, when we treat patients with phenylketonuria by means of an appropriate diet from birth, we can prevent both physical and mental disease; a striking example of how Nurture can modify Nature in the clinic, as well. Clues from the study of depression suggest that the family serpent of gloom in the blood may also turn out to be an inborn error of metabolism. The irony would not have escaped either William or Alice James that the Garrods, who began the science of gout, were present at the birth of "neurotic science."

A last irony in the story of Alice James is that her legacy is one she could not have imagined. Far better known today than in her lifetime, she has been variously celebrated as a model memoirist, a feminist critic of Victorian mores, a pioneer of narrative medicine, and a lesbian heroine. Over the years, she has become an exemplar in arguments over the role that male physicians play in imposing diseases on women, the role of gender in renown, and the requirement for victimhood in literary honor. Her journal fits into a genre Elaine Showalter has called *Hysteries* (1998).[37] Cathleen Schine's first novel *Alice in Bed* (1983) is about a woman hospitalized for a medically unexplained disease; the doctors are demons.[38] Susan Sontag's more recent play, also entitled *Alice in Bed* (1993), is a more fitting tribute; it begins like Gertrude Stein and ends in a benediction.[39] The play opens with the voice of her nurse in Leamington:

NURSE: Of course you can get up
ALICE: I can't
NURSE: Won't
ALICE: Can't
NURSE: Won't

Later Alice soliloquizes about Rome, a city that she, unlike her famous brothers, has never seen: "I would be very humble. Who am I, compared with Rome. I come to see Rome, it doesn't come to see me. It can't move." The play ends with Alice saying, "Let me fall asleep. Let me wake up. Let me fall asleep." To which her nurse replies, "You will."

She wakes up on every page of her journal.

Lewis Thomas and the Two Cultures

Bien écrire, c'est tout à la fois bien penser, bien sentir et bien rendre; c'est avoir en même temps de l'esprit, de l'âme et du goût. Le style suppose la réunion et l'exercice de toutes les facultés intellectuelles.

To write well is at once to think, feel and express oneself well; simultaneously to possess wit, soul and taste. Style comes from the integration and exercise of all the intellectual faculties. —Comte de Buffon.[1]

I REMEMBER THE MOMENT THAT Lewis Thomas asked me to be his first chief resident in the old redbrick tenement of Bellevue Hospital. He had a reputation of charming young doctors into academic medicine under conditions and for wages that few dockworkers would tolerate. At the time I was still deciding on whether to follow my father into practice or move on to an academic career. He told me that his chief resident wouldn't have an office, but a lab. He'd be off every third night and the job would be the first step on the academic ladder. Then Thomas told me what academic salaries were like in the late 1950's, and my rude, younger self quoted:

"What is science but the absence of prejudice backed by the presence of money?"

"Henry James" Lewis Thomas snapped, "from *The Golden Bowl*, Chapter One." He went on,

"All right then, you won't earn very much but you'll have a lot of fun in the lab and time to read. If you're lucky, you may also discover something. It's a great life."

The life of Lewis Thomas spanned the golden age of American medicine, an era when—in his words—our oldest art became the youngest science. Thomas played a major role in that transformation; he was

Dr. Lewis Thomas

known among scientists as an innovative immunologist, pathologist, and medical educator. He became far better known as a deft writer whose essays bridged the two cultures by turning the news of natural science into serious literature. Witty, urbane, and skeptical, he may have been the only member of the National Academy of Sciences to have won both a National Book Award and an Albert Lasker Award. He is certainly the only medical school dean whose name survives on professorships at Harvard and Cornell, a prize at Rockefeller University, a laboratory at Princeton—and on a book that is eleventh on the Modern Library's list of the best 100 nonfiction books of the 20th century.

Thomas made three important discoveries in the field of which he was a pioneer, immunopathology; each had implications for human disease. He found that the white cells of blood, the leukocytes, were important mediators of fever and shock brought about by bacterial endotoxins; this taught us how microbes kill us if we don't first kill them. He also made the novel observation that proteolytic enzymes such as papain could injure cartilage when injected into the circulation, the same sort of damage results when our own cells release papainlike ferments; this line of investigation showed us how joints destroy themselves in arthritis. But perhaps his most prescient suggestion, made years before the HIV pandemic, was that our immune system constantly surveys our body to find and destroy aberrant cancer-prone cells; we now attribute Kaposi's sarcoma and other AIDS-related tumors to defects in Thomas's "immune surveillance." Those discoveries were made in a very intense period of bench research (1950–1965) at the University of Minnesota and at NYU before he turned his attention to the broader issues of science and to his writing.

The lifetime of Lewis Thomas coincided with a special period in American medicine, a time when its scientific base became the strongest it had ever been and its social impact the greatest. Indeed, judging from the numbers who came from overseas to learn from it, American medicine became the envy of the world. That shift of balance from the old world to the new happened at the same time that doctors dropped "the laying on of hands" and took up the task of monitoring machines. It was not by accident but by design that American medicine was turned from a nineteenth-century folk art into a twentieth-century science. After the Flexner report of 1910, medical instruction became largely concentrated in university hospitals where the modern sciences of immunology, biochemistry, and genetics could be pursued as eagerly at the bedside as in the lab. Lewis Thomas and his generation of immunologists presided over the conquest of polio and rheumatic fever; the achievements of blood-banking, cardiac surgery, and the transplantation of organs; not to speak of the discovery that DNA was the basic unit of genetic information. In the words of C. P. Snow, they had the future in their bones.[2] Like Snow himself, they were, in the main, committed skeptics.

The Education of Lewis Thomas

Lewis Thomas grew up as a bright lad in a loving family in a comfortable house in Flushing, Queens. His father, Dr. Joseph Simon Thomas (Princeton, 1899, Columbia P&S, 1904), was a good-natured, hard-working doctor who had met and married the love of his life, Grace Emma Peck of Beacon Falls, Connecticut, at Roosevelt Hospital where she was a nurse and he was an intern. They were married in New York City on the 30th of October 1906 and thereafter, in the words of her son, Emma Peck's nursing skills were "devoted almost exclusively to the family."[3]

Lewis Thomas was born on November 25, 1913. As were his three older sisters and younger brother, Lewis was sent to the local schools. But soon the family decided that Flushing High School was not quite ready to prep another Thomas for Princeton. After three semesters in Queens, Lewis Thomas transferred to the McBurney School, a less than exclusive prep school in Manhattan. He graduated in 1929 in the top quarter of his class. Medical practice was to protect the Thomas family against the worst of the Great Depression, which began on Black Tuesday, exactly one month after fifteen-year-old Lewis left for Princeton, in September of 1929.

At Princeton he "turned into a moult of dullness and laziness, average or below in the courses requiring real work." He took little interest in physics or inorganic chemistry and dismissed athletics as a general waste of effort. By reason of youth and family standing, he ranked low in the eating club hierarchy of prewar Princeton and was grateful to find safe haven at Key and Seal, a club that was, literally, the furthest out on Prospect Avenue. But high spirits and natural wit brought him to the offices of the *Princeton Tiger* where Thomas soon published satires, poems, and parodies under the nom de plume of ELTIE. "After the crash of '29, we were in thrall to Michael Arlen; we slouched around in Oxford bags and drank bootleg gin from the tub like Scott and Zelda" Thomas recalled. "They told us we'd go out like a light from that stuff. Out like a light. I think I did a piece on bootleg gin for the *Tiger* about that." He had; it's unreadable. Then, on one winter weekend visit to Vassar in 1932, Lewis Thomas met a young freshman from Forest Hills. Her father was a diplomat, her name was Beryl Dawson, and after years of separation for one or another reason they were married a decade later. By then the moult had spread its wings.

Years later, his editor, Elisabeth Sifton, asked me "When was it that Thomas became so wise?" Thomas attributed his metamorphosis to his senior year at Princeton and a biology course with Professor Wilbur Swingle. Swingle's discovery of a life-saving adrenal cortical extract—a crude version of deoxycorticosterone—had won wide acclaim. Thomas recalled that Swingle sparked his lifelong interest in the adrenals. Swingle also introduced him to Jacques Loeb's literary/philosophical speculations on ions and cell "irritability" in *The Mechanistic Conception of Life*. Five years out of Princeton, young Thomas would sign up to work with Jacques Loeb's son, Robert F. Loeb, at Columbia.

In his senior year, the depression hit home and Thomas knew that getting into medical school was one solution to the unemployment problem. He also confessed he had a leg up on other applicants:

> I got into Harvard . . . by luck and also, I suspect, by pull. Hans Zinsser, the professor of bacteriology, had interned with my father at Roosevelt and

> had admired my mother, and when I went to Boston to be interviewed in the winter of 1933 [Zinsser] informed me that my father and mother were good friends of his, and if I wanted to come to Harvard he would try to help. . . .[3]

Help he did and Thomas entered Harvard at the age of 19 in the fall of 1933. Thomas's career at the Harvard Medical School turned out just fine; he received grades far better than at Princeton. When asked in 1983 which member of the Harvard faculty had the greatest influence on his medical education, Thomas replied, "I no longer grope for a name on that distinguished roster. What I remember now, from this distance, is the influence of my classmates."[3] Nevertheless, some on that roster made a lasting impression. Hans Zinsser[4] in Bacteriology showed that it was possible to function both as a laboratory scientist and a respected writer; Walter B. Cannon in Physiology taught him that the details of homeostasis held the keys to *The Wisdom of the Body*; David Rioch in Neuroanatomy had him build a wire and plasticene model of the brain which Thomas trekked about for fifteen years; and in Tracy Mallory's office Lewis Thomas came across a pickled specimen which, "like King Charles's head," would haunt his investigative career for decades to come.[4]

At one of Mallory's weekly pathology seminars in the depths of Massachusetts General Hospital, Thomas leaned back in his chair and by accident knocked over a sealed glass jar containing the kidneys of a woman who had died of eclampsia. Replacing the jug, he noted that both organs were symmetrically scarred by the deep, black, telltale marks of bilateral renal cortical necrosis. Thomas remembered having seen something like those pockmarked kidneys before. They had been provoked in rabbits by two appropriately spaced intravenous injections of endotoxin: it was called the generalized Shwartzman phenomenon and he would tussle with it for the rest of his scientific career.

Thomas graduated cum laude from the Harvard Medical School in 1937 and began internship on the Harvard Medical Service of the Boston City Hospital. A history of the Harvard Medical Unit at Boston City Hospital (see "Mother of Us All") documents that of the 71 young physicians who trained there between 1936 and 1940, 52 became professors of medicine, while 6 went on to the deanship of medical schools.

He remained at Boston City until 1939, when the confluence of his interests in neurology, adrenal hormones, and the Loeb mystique brought him to New York. Halfway through his internship in Boston, Thomas heard that Dr. Robert F. Loeb was becoming director of the Neurological Institute in New York and resolved to study with him because

> Loeb was a youngish and already famous member of the medical faculty in the Department of Medicine at P&S, recognized internationally for his work on Addison's disease [and] the metabolic functions of the adrenal cortex and the new field of salt and water control in physiology.[3]

He served as a neurology resident (his only specialty training) and research fellow at P&S from 1939 to 1941 with time out to marry Beryl at Grace Church in New York in January of 1941. Robert Loeb abruptly moved to the chairmanship of Medicine but Thomas found that there was a fellowship with John Dingle awaiting him back at Harvard and jumped at the chance.

Almost as soon Lew and Beryl had established themselves back in Boston, Thomas was sent by Dingle on a month-long medical mission to Halifax, where an outbreak of meningococcal meningitis had struck the wartime port. Beryl served as lab assistant. Those four weeks in the field, an important publication on the effects of sulfadiazine in meningitis, and a thorough grounding in immunology in Dingle's lab, were a prelude to a naval commission after Pearl Harbor. Thomas reported in March 1942 to the Naval Research Unit at the Hospital of the Rockefeller Institute and on January 12, 1945, landed with a detachment of that Unit headquartered on Guam. Thomas and Horace Hodes were put to work on Japanese B encephalitis on Okinawa, and quickly identified horse blood as a reservoir for the virus.

War ended, waiting to be sent stateside, Thomas began experiments on rheumatic fever on Guam. Putting unused lab facilities to good purpose, he knew that rheumatic fever was almost always preceded by a streptococcal throat infection and that the interval between infection and heart disease could be very long indeed. Perhaps the disease was an allergic reaction to the microbe, to the patients's own tissues or to a mixture of the two.

On Guam, Thomas found that rabbits receiving a mixture of microbes (streptococci) and ground-up heart tissue became ill and died within two weeks; the microscope revealed that their hearts showed lesions that resembled those of rheumatic fever in humans. Control rabbits injected with streptococci alone or with ground-up heart tissue alone remained healthy and showed no such damage. Thomas was entirely confident that he had solved the whole problem of rheumatic fever. He hadn't. On his return to the Rockefeller Institute in New York, he couldn't repeat those experiments, sacrificing "hundreds of rabbits, varying the dose of streptococci and heart tissue in every way possible." He was vastly relieved that he hadn't rushed into print on the basis of those rabbits on Guam.

Thomas's first faculty position after discharge from the navy in 1946 was as assistant professor of pediatrics at the Harriet Lane Home for Invalid Children of Johns Hopkins. Thomas tried once more to repeat those rabbit experiments. This time, he mixed streptococci and heart tissue with a devilish brew of fats (Freund's adjuvants) that had produced tissue injury in other disease models. Bad news for assistant professor Thomas: the rheumatic fever experiments failed once again. But Thomas could not shake off those experiments that had worked so well on Guam. Perhaps the host, the rabbits in Guam, for example, but not those in New York or Baltimore, had been "prepared" by an earlier insult as by that endotoxin injection in the Shwartzman phenomenon.

He tackled the problem with Chandler (Al) Stetson, a lifelong friend who was to become his colleague in Minnesota, and his successor as the Professor of Pathology at NYU. Thomas and Stetson "prepared" rabbits with endotoxin from meningococci. The prepared skin had an excess of acid and they reasoned that the acid might activate the tissue's own ferments, proteases called cathepsins. But they were neither able to measure cathepsin activity, nor obtain purified cathepsins, so they next injected rabbits with enzymes obtained off the shelf, trypsin and papain to be exact. Trypsin was ineffective, but papain produced lesions in the skin that looked very much like the local Shwartzman reaction.

When Thomas left Hopkins, he took the problem with him. He served a brief stint at Tulane where he became a Professor of Medicine and Director of the Division of Infectious Disease. He was diverted for a while by studies of circulating antibodies in animal models of multiple sclerosis, but returned to rheumatic fever when he was appointed as American Legion Professor of Pediatrics and Medicine at the University of Minnesota in 1951. In quick time, he put together a team of young investigators, most of whom were soon at work on the Shwartzman phenomenon and the streptococcus.

He reverted to the notion that proteases, either secreted by the streptococcal microbe or released from the victims's own cells, caused damage in a "prepared" heart or joint. With a young Minnesota pediatrician, Robert A. Good ("the smartest investigator I ever met," he once told me), he found out that if one removed white cells from the Shwartzman equation kidney injury was prevented. The kidneys were also spared if one gave heparin, which prevented blood vessels from becoming plugged by fibrin, platelets, and white cells. Good and Thomas suggested that "a combination of humoral and cellular factors made by the host caused the tissue injury." Nowadays we invoke complicated systems with complicated names such as anaphylatoxins, Toll receptors, apoptosis, caspases, and cytokines to explain the Shwartzman phenomenon. But in the 1950's Good and Thomas had provided a satisfactory explanation and the flow of satisfying, explanatory papers followed Thomas from Minnesota as he moved to NYU in 1954.

Thomas was recruited to NYU by Colin McLeod to become Professor and Chairman of the Department of Pathology. He was delighted to return to the metropolis with Beryl and his three daughters, Abigail (b. 1941), Judith (b. 1944), and Elizabeth (b. 1948), and to set up his household at Sneden's Landing, a small town up the Hudson from the city. He remained at NYU for fifteen years and proceeded to turn it into a world center of immunology, first in Pathology (1954–1958), then as Professor and chairman of the Department of Medicine at NYU–Bellevue Medical Center. Director of III and IV Medical Divisions, Bellevue Hospital (1958–1966) and finally as Dean of the New York University School of Medicine and deputy director of NYU Medical Center (1966–1969).

Over those years he attracted and/or trained a legion of scientific stars and superstars at NYU: Frederick Becker, Baruj Benacerraf, John David, Edward Franklin, and Emil Gottschlich, Howard Green, H. Sherwood Lawrence, Robert T. McCluskey, Peter Miescher, Victor and Ruth Nussenzweig, Zoltan Ovary, Jeanette Thorbecke, Stuart Schlossman, Chandler Stetson, Jonathan Uhr, and Dorothea Zucker-Franklin. Thomas's international colleagues were frequent visitors: Sir Macfarlane Burnett, Dame Honor Fell, Philip Gell, James Gowans, Sir Peter Medawar, Thomas Sterzl, and Guy Voisin.

Early on in his NYU days Thomas hit a rough patch. Whereas cortisone, the miracle drug, clearly stopped inflammation in the clinic, Thomas was astonished to find that cortisone not only proved ineffective against the Shwartzman phenomenon, but actually provoked it. This puzzle took the wind out of his sails. He was indeed "in irons on his other experiments" and "not being brilliant." Then came the floppy-eared bunnies (see "Floppy-eared Rabbits").

After papain, new discoveries proceeded apace. If an exogenous protease caused connective tissue damage, where might endogenous proteases reside? Thomas spent a summer with Dame Honor Fell, director of the Strangeways Research Laboratory in Cambridge. Fell had been studying vitamin A and had found that it produced depletion of cartilage matrix in mouse bone rudiments growing in a dish. Fell and Thomas decided to trade experimental systems. They first added papain to the little bone cultures in the dish and were able to produce vitamin A–like lesions in mouse cartilage. Thomas then returned to NYU to do the reciprocal experiment. Together with Jack Potter and R. T. McCluskey, Thomas and I stoked rabbits full of the vitamin A and sure enough: twenty-four to forty-eight hours later, their ears drooped as if they had been given papain. We were convinced then that Vitamin A in some fashion released an endogenous papainlike enzyme from cartilage cells and that this enzyme proceeded to break down cartilage matrix. At the time we supposed that the enzyme was present in lysosomes, subcellular "suicide sacs" that had just been described by Christian de Duve at Louvain. We suggested that vitamin A had ruptured the walls around these organelles, and that cortisone and its analogues must therefore stabilize the lysosomes.

These days the answer is more complicated. Nowadays, we believe that metalloproteinases are released from cells and that synthesis of these proteases is under opposing transcriptional control by vitamin A and cortisone acting via well-defined cytoplasmic and nuclear receptors. Cortisol receptors recognize palindromes of DNA; vitamin A receptors see tandem response elements of DNA; there are at least two types of glucocorticoid receptors, these antagonize fos/jun transcription factors, and so on . . . and so on . . . in abundant detail. It all seemed simpler a generation ago. But these experiments, the last in which Thomas played a hands-on role, pointed the way for Thomas's students and, in turn, *their* students to explore other areas of human biology: how infection and immunity make our white cells clump and stick to blood vessels; how stimulated white cells release molecules (cytokines) that cause fever, fatigue, and inflammation; how tissue are recognized as foreign and transplants are rejected; how cortisone and aspirin-like drugs work in arthritis—and, as a follow-up of the cortisone/lysosome experiments, how to design, manufacture, and bring to the clinic, tiny drug-bearing lipid structures called liposomes, that have saved thousands of lives.

After 1965, Thomas moved from the lab bench to the rougher terrain of medical administration and science policy. Thomas had a broad interest in how medical science shapes, and is shaped by, society. Wit, candor, and attention to principle rather than politics made him a valuable spokesman for medical science. While still at NYU, Thomas served as a member of the New York City Board of Health (1957–1969), was instrumental in the construction of the new Bellevue Hospital, and set up the Health Research Council, a sort of local NIH. As chairman of the Narcotics Advisory Committee of the New York City Health Research Council, he guided Vincent P. Dole into methadone research and pointed Eric Simon to endorphins (1961–1963). After a stint in New Haven as Professor of Pathology and Dean (1969–1973) at Yale University School of Medicine, he became president and chief executive officer of the Memorial Sloan-Kettering Cancer Center (1973–1980). At MSKCC, he launched a major attack on tumor immunology, recruiting Robert Good as director; Thomas became Chancellor of MSKCC from 1980–1983. In retirement, his summer home in the Hamptons made a University Professorship at SUNY–Stony Brook (1984) convenient and his Manhattan apartment let him serve as Writer in Residence at the Cornell University Medical School until his death in 1993. By then he had been for several decades the most widely read interlocutor between the older literary culture and the new world of medical science.

Lewis Thomas: the Element of Style

Lewis Thomas was preceded in his role as medical scribe to the nation by such other American physician/writers as Oliver Wendell Holmes, William James, Walter B. Cannon, and Hans Zinsser. Each contributed vastly to the biology of medicine, each wrote books that gained a broad general audience, each taught that science is a very human activity. Zinsser, who as we've seen, was responsible for Thomas's admission to the Harvard Medical School, summed up the tradition by proposing that:

> Aside from the habits of hard work that [medicine] demands, it embraces a broad survey of the biological field, enforces a considered correlation of the fundamental sciences, and, on the human side, brings the thoughtful student face to face with the emotional struggles, the misery, courage and cowardice of his fellow creatures.[4]

Like Zinsser before him, Thomas fused the two cultures of "fundamental science" and "the human side" because for him they were one.

He went on to become a persuasive spokesman for the biological revolution because he was convinced that while science is only one of the many ways we have of making sense of nature, medical science is the only way we have of making sense of disease. That conviction would have remained more or less private were it not for the public language in which it was voiced. His sound was distinct and unmistakable, the prose direct and limpid. Here, for example, is Thomas's suggestion for signals we might send from earth to announce ourselves to whatever life there might be in outer space:

> Perhaps the safest thing to do at the outset, if technology permits, is to send music. This language may be the best we have for explaining what we are like to others in space, with least ambiguity. I would vote for Bach, all of Bach, streamed out into space, over and over again. We would be bragging, of course, but it is surely excusable for us to put the best possible face on at the beginning of such an acquaintance. We can tell the harder truths later. . . . Perhaps, if technology can be adapted to it, we should send some paintings. Nothing would better describe what this place is like, to an outsider, than the Cézanne demonstrations that an apple is really part fruit, part earth.[5]

That sort of writing is the product of another unique period in American culture. Thomas and his colleagues were educated in colleges at which the liberal arts were still firmly in place and John Dewey's learning-by-doing had moved from primary schools into the universities. It was an era when those who did medical science were expected to know why it was done and for whom. They were also expected to make only modest claims for their success: "I was lucky," Thomas quipped after he received a medal at Bologna in 1978, "chance favored the prepared grind." One knew that he was speaking for a generation of medical scientists who believed that one could do serious work without taking oneself too seriously.

Thomas wrote one, rather impersonal, memoir, *The Youngest Science*.[3] Thomas believed that a scientific memoir ought to remember not only how the science came about but how it felt at the time: sometimes right, sometimes wrong, always surprising. Thomas called it a story of finding a pattern in a jumble, a task even he found daunting:

> It should be easier, certainly shorter work to compose a memoir than an autobiography, and surely it is easier to sit and listen to the one than to the other. . . . In my [memoir] I find most of what I've got left are not memories of my own experience, but mainly the remembrance of other people's thoughts, things I've read or been told, metamemories. A surprising number

> of them turn out to be wishes rather than recollections . . . hankerings that the one thing leading to another had a direction of some kind, and a hope for a pattern from the jumble—an epiphany out of entropy.[6]

That passage, precise and informal at once, illustrates the flow of Thomas's thought and speech. Thomas was as likely on the wards as in print to pair epiphany (à la James Joyce) with entropy (à la the second law of thermodynamics, or $\Delta S > q/T$). That balance of phrase could be said to be the signature of Thomas's prose; epiphany seemed to be having it out with entropy on every page. He was also sparing of words when fewer spoke louder. Half a year before his death from Waldenström's macroglobulinemia, a slow, wasting form of bone marrow disease, he received an award named in his honor at Rockefeller University. Confined to a wheelchair by his illness, he declined the podium and apologized to the audience for "not rising to the occasion." About the same time, I had reached him on the telephone:

"How are you doing?" I asked. He knew what I was asking.

"So," he replied.

"What do you mean by 'so'?"

"Well," said Thomas, "in my family, there were only three ways of answering that question of yours. If things were going along splendidly, you'd answer 'fine.' If there were a bit of trouble around, you'd say 'so-so.' Right now, I'm 'so.'"

When more words were required, they flowed like wine. Lewis Thomas's chosen means of expression was the informal essay, a literary form that accommodates many topics but always has the mind of its author as the subject. A reader of his pieces quickly becomes aware that Thomas has invited him to a tug of war between two turns of mind, a playful match between two equally attractive personae: a cheerful Thomas who urged holism and a doubting Thomas who was a card-carrying reductionist. Holism, as I learned from Lewis Thomas, was invented by General Jan Smuts—the intelligent design chap—in 1926.[7] It implies that Matter and Life are one. Reductionism, as we've seen, derives from Hippolyte Taine who borrowed the term from the chemists who use it to denote an agent that reduces a compound to a simpler substance by removing oxygen. Matter without Life, one might say.

As a medical scientist, Thomas was persuaded that only patience, doubt and diligence, the reductionist virtues, could pluck facts from nature. But Thomas also understood the very human need to turn the strands of fact into a fabric of belief. In that mode he had but one exemplar: William James. Especially in his later, more ruminative essays Thomas successfully blended the Jamesian "Will to Believe" with James Lovelock's newer Gaia hypothesis, a postulate that life on our planet has been chiefly responsible for the regulation of that life's own environment.[8] Lovelock's holistic notion seems to unite the best of James with the best of Thomas and it is no mean compliment to suggest that a passage such as this from James would fit comfortably in any of Thomas's essays:

> We find ourselves believing, we hardly know how or why. We all of us believe in molecules and the conservation of energy, in democracy and necessary progress...all for no reason worthy of the name. We see into these matters with not more inner clearness, and probably with much less, than any disbeliever in them might possess. . . . For us not insight, but the prestige of the opinions, is what makes the spark shoot from them, that lights up our sleeping magazines of faith. Our faith is faith in someone's else's faith, and in the greatest matters this is most the case. Our belief in truth itself is that there is a truth and that our minds and it are made for each other.[9]

Like James, Thomas was celebrated as a "poetic" or "creative" writer and scientist. But modern critics use those adjectives in much the way that eighteenth-century essayists would have used "sentimental." Thomas was by no means a sentimental essayist. I'm convinced that there was more structure than sentiment to his writing, just as there was more science to his art than art to his science. Here is a passage with a Jamesian sense of our planet:

> The overwhelming astonishment, the queerest structure we know about so far in the whole universe, the greatest of all cosmological scientific puzzles, confounding all our efforts to comprehend it is the earth. We are only now beginning to appreciate how strange and splendid it is, how it catches the breath, the loveliest object afloat around the sun, enclosed in its own blue bubble of atmosphere, manufacturing and breathing its own oxygen, fixing its own nitrogen from the air into its own soil, generating its own weather at the surface of its own rain forests, constructing its own carapace from living parts: chalk cliffs, coral reefs, old fossils from earlier forms of life now covered by layers of new life meshed together around the globe, Troy upon Troy.[10]

Again like James, Thomas was not simply a clever scientist with a creative turn of mind; he was a writer to the bone. Evelyn Waugh, whom Thomas had admired since his undergraduate days on the *Princeton Tiger*, introduced the term "architectural" to describe the difference:

> Creative is an invidious term [for a writer] . . . a better word, except that it would always involve explanation, would be "architectural." I believe that what makes a writer, as distinct from a clever and cultured man who can write, is an added energy and breadth of vision which enables him to conceive and complete a structure.[11]

The architectural structure that Thomas worked out fitted readily into the conventions of the informal essay. He derived from the facts of natural science, such as the workings of inflammation, symbiosis, the planetary ecosystem, or the life of social insects, metaphors for broad aspects of human activity, such as curiosity, language, or altruism. Fact marched hand-in-hand with solace; he assured us that a meningococcus with the bad luck to catch a human was in more trouble than a human who catches a meningococcus. Who on earth would not welcome those tidings of comfort and joy? Every once in a while Thomas reversed the direction of his metaphors, using human behavior or language as a metaphor for the odd fact of cell biology:

> The meaning of these stories [of protozoan symbiosis] may be basically the same as the meaning of a medieval bestiary. There is a tendency for living things to join up, establish linkages, live inside each other, return to earlier arrangements, get along, wherever possible. This is the way of the world.[12]

But his years in the lab served him well on the page. His sense of trial and error at the bench and in the clinic, of how cells divide, microbes hurt, and creatures die, gave a tough edge to his writing:

> When injected into the bloodstream, [endotoxin] conveys propaganda, announcing that typhoid bacilli (or other related bacteria) are on the scene and a number of defense mechanisms are automatically switched on, all at once . . . including fever, malaise, hemorrhage, shock, coma and death. It is something like an explosion in a munitions factory.[13]

Thomas had been writing for publication since his efforts on the *Princeton Tiger*. His poetry became more ambitious and in his house office days he published several remarkably polished verses in literary magazines such as the *Atlantic Monthly*. One of these even appeared in 1944 while he was in the Pacific with the Rockefeller Institute Medical Research Unit.[14] But, for twenty years after Thomas returned from the war, his writing was pretty much limited to the scientific literature. Later, after he had made his contributions to immunology, after he had secured his reputation in science, and while he was serving as chairman, dean, and chancellor at three sometimes exasperating institutions, he turned his attention once more to the muse.

In 1965, he had permitted himself a ruminative essay on inflammation which was brought to the attention of Franz Ingelfinger, editor of the *New England Journal of Medicine* and an old colleague of Thomas's at Harvard. At Ingelfinger's request, Thomas began his stint as author of the bimonthly column "Notes of a Biology Watcher." Thanks to Elisabeth Sifton, then an editor at Viking Press, those sparkling essays were soon collected into *The Lives of a Cell*, the volume became a best seller, won a National Book Award—and Thomas was well on his way to a place in the world of letters.

Unlike other scientist/writers who tend to limit their subjects to their own field of research—evolution is the core of Stephen Jay Gould, for example—Thomas worked hard at the task of writing well about all manner of things. Waugh would have appreciated that effort:

> In youth high spirits can carry one over a book or two. The world is full of discoveries that demand expression. Later a writer must face the choice of becoming an artist or prophet. He can shut himself up at his desk and selfishly seek pleasure in the perfecting of his own skill or he can pace about, dictating dooms and exhortations on the topics of the day. The recluse at the desk has a bare chance of giving abiding pleasure to others; the publicist has none at all.[15]

Tough critics of his science claimed that much of Thomas's immunology was small-scale and anecdotal. But Thomas's lasting contributions to inflammation and immunity can be readily identified today, his notions are imbedded in the history of immunology.[16] On the literary side, Stephen Jay Gould accused his essays of disguising their serious themes as "homegrown Yankee wisdom" cloaked in "charming and superficially rambling" accounts—too charming, perhaps, for words.[17] Sidney Hook complained about his memoir that "Mr. [*sic*] Thomas's 'A Long Line of Cells' is an instructive lesson in biology but tells us nothing about his life that distinguishes it from any other human life."[18] Christopher Lehmann-Haupt accused Thomas of "Optimism (relentless) on humanity."[19] Thomas himself confessed that he may have told us once too often that all is for the best in this best of all possible worlds: "I'm not sure Pangloss was all that wrongheaded. This is in real life the best of all possible worlds, provided you give italics to that word possible."[6] But his mandarin wit prevailed over the Panglossian strain; he became a fine essayist by the best means available, which was to work hard at writing well. Indeed, his very last, somewhat slight, book, *Etcetera, Etcetera* was devoted entirely to words and the sound of words. That attention to the *mot juste* reminds one of Waugh's advice that over time the writer is better off perfecting his style rather than peddling his subject:

> Literature is the right use of language irrespective of the subject or reason of utterance. A political speech may be, and often is, literature. A sonnet to the moon may be, and often is, trash. Style is what distinguishes literature from trash. . . . The necessary elements of style are lucidity, elegance, individuality; these three qualities combine to form a preservative which ensures the nearest approximation to permanence in the fugitive art of letters.[20]

That point taken, I'd make the partisan argument that Thomas' careful attention to style, which so clearly meets Waugh's criteria of lucidity, elegance, and individuality, gives Lewis Thomas a shot at permanence in the world of letters. A number of his compositions stand up to essays by such other modern masters of the genre as E. B. White, A. J. Liebling, and John Updike. In a select few, Thomas reaches back to touch the mantle of Montaigne.

Rats, Lice, and Zinsser

LIKE MANY OF MY COLLEAGUES in academic medicine, I caught my first whiff of science from popular books about men and microbes. By the time we had finished high school, most of us had read, and often reread, Paul de Kruif's *Microbe Hunters*, Sinclair Lewis's *Arrowsmith*, and *Rats, Lice and History* by Hans Zinsser. It's hard, nowadays, to reread the work of de Kruif or Sinclair Lewis without a chuckle or two over their quaint locution, but Zinsser's *raffiné* account of lice and men remains a delight. Written in 1935 as a latter-day variation on Laurence Sterne's *The Life and Opinions of Tristam Shandy*, Zinsser's book gives a picaresque account of how the history of the world has been shaped by epidemics of louse-borne typhus. He sounded a tocsin for the war on microbes in the days before antibiotics; his challenge remains meaningful today:

> Infectious disease is one of the few genuine adventures left in the world. The dragons are all dead and the lance grows rusty in the chimney corner. . . . About the only sporting proposition that remains unimpaired by the relentless domestication of a once free-living human species is the war against those ferocious little fellow creatures, which lurk in dark corners and stalk us in the bodies of rats, mice and all kinds of domestic animals; which fly and crawl with the insects, and waylay us in our food and drink and even in our love.[1]

Despite an unwieldy subtitle, à la Sterne, *"Being a study in biography, which, after twelve preliminary chapters indispensable for the preparation of the lay reader, deals with the life history of TYPHUS FEVER," Rats, Lice and History* became an international critical and commercial success. Zinsser's romp through the ancient and modern worlds describes how epidemics devastated the Byzantines under Justinian, put Charles V atop the Holy Roman Empire, stopped the Turks at the Carpathians, and turned Napoleon's Grand Armée back from Moscow. He explains how the louse, the ubiquitous vector of typhus, was for most of human history an inevitable part of existence, "like baptism, or smallpox"; its habitat extended from hovel to throne. And after that Murder in the Cathedral, the vectors deserted Thomas à Becket:

> The archbishop was murdered in Canterbury Cathedral on the evening of the twenty-ninth of December [1170]. The body lay in the Cathedral all night, and was prepared for burial on the following day . . . He had on a large brown mantle; under it, a white surplice; below that, a lamb's-wool coat; then another woolen coat; and a third woolen coat below this; under this, there was the black, cowled robe of the Benedictine Order; under this, a shirt; and next to the body a curious hair-cloth, covered with linen. As the body grew cold, the vermin that were living in this multiple covering started to crawl out, and, as . . . the chronicler quoted: 'The vermin boiled over like water in a simmering cauldron, and the onlookers burst into alternate weeping and laughter . . .'[2]

Zinsser's literary range and magpie intellect prompted reviewers of the day to compare him to an earlier Harvard Medical School author, Dr. Oliver Wendell Holmes. Indeed, the first few chapters of *Rats, Lice and History* sound a lot like Holmes's *The Autocrat of the Breakfast Table*. Both works offer clean draughts of political and poetic history "dipped from the running stream of consciousness"—to use Holmes's phrase, later made famous by William James. *Rats, Lice and History* is still in print; but it's playing to a different audience these days. A Yale academic worries that readers "may be puzzled by the acerbic references to the literary dandies of the interwar period,"[3] while a scroll down the *Amazon.com* Web site yields another complaint "Some of the writing assumes that all readers were educated under an aristocratic university system, so that there are bits thrown in in Latin and Greek, not to mention French and other modern languages."[4] Be that as it may, Zinsser's assumption of an "aristocratic university system" did not prevent the book from becoming a best seller in 1935, nor to undergo seventy-five subsequent printings. And as for those literary dandies: T. S. Eliot? Gertrude Stein? Lewis Mumford? Edmund Wilson?

Zinsser followed up with *As I Remember Him: The Biography of R.S.*, a third-person autobiography that survives as a distinguished work of literature. The R.S. of Zinsser's title is an abbreviation of the "Romantic Self," or the last letters, inverted, of Hans Zinsser's first and last names. In it, Zinsser spells out his warm view of medicine as a learned profession:

> There is in it a balanced education of the mind and of the spirit which, in those strong enough to take it, hardens the intellect and deepens the sympathy for human suffering and misfortune. . . .[5]

As I Remember Him was written two years before Zinsser's death of lymphatic leukemia in 1940 at the age of sixty-one. A selection of the Book-of-the-Month Club, it had reached the best-seller list as its author lay dying; news of its warm reception by book reviewers filtered into the obituaries. The volume was widely popular among doctors of my father's generation. The copy I first read had sat on one of my dad's office shelves, wedged between Axel Munthe's *The Story of San Michele* and Romain Rolland's *Jean-Christophe*.

As I Remember Him tells the story of Zinsser and his cohort of American physicians who "were more fortunate than they knew, because they were about to participate in a professional evolution with few parallels." Zinsser lived long enough to see American medicine develop from a "relatively primitive dependence upon European thought to its present magnificent vigor."[6] He describes how over the years the torch of medical science passed from England to France to Germany. The flow of American medical students followed its path, which had led in Central Europe to "a powerful reaction of the basic sciences upon medical training and a true spirit of research [that] pervaded medical laboratories and clinics."[7] The Golden Age of German medicine ended abruptly in 1933 when, as Zinsser lamented, "common sense became counter-revolution."

Zinsser, a New Yorker, came of German liberal stock of the generation of 1848. "The spirit of this age of my German grandfathers was one of growing philosophical materialism and the *Freie Deutsche Gemeinde* (free German communities) or 'ethical culture societies' of the sort carried on for many years by Felix Adler."[8] Zinsser's father, a free-thinking industrial chemist, raised him in a cultivated Westchester home, with trips abroad, private tutors, music and riding lessons. He prepped for Columbia at a small private school on Fifty-ninth Street run on Ethical Culture lines by Dr. Julius Sachs (now the Dwight School). At Columbia College, he fell under the spell of George Woodberry, the Mark van Doren of his day, and became enchanted by the poet and essayist who had had been hailed as "the American Shelley" by James Russell Lowell. Woodberry not only introduced Zinsser to the New England Enlightenment, but also sparked his lifelong love affair with poetry.

Zinsser's Columbia years reinforced his distrust of credulity, later strengthened by the skepticism of the German Enlightenment as interpreted by H. L. Mencken. Heine, the skeptic was the Zinsser family hero, and Zinsser often repeated the poet's final words. Asked to turn to God, that He might pardon him at death, Heine replied, "God will forgive me. It's his business." It was in this spirit that Zinsser differed from William James.

> I have been utterly incapable of that "over-belief" which William James postulates as necessary to faith. Moreover, to give religious experience—as he does—a merely pragmatic value seems both to be begging the question and to be making light of a grave problem.[9]

At Columbia, Zinsser may have fallen under the spell of George Woodberry in the humanities, but his mentor in science was Edmund B. Wilson (no relation to the writer). Zinsser vastly admired Wilson, a founder of developmental genetics in America. Wilson and his assistants, Zinsser knew, were the direct spiritual offspring of the Darwinian period; they had known Haeckel and Huxley. Wilson himself had described how chromosomes divide before cell cleavage and are united in pairs in the new cells. Zinsser became a convert to experimental biology after many hours in Wilson's cytology labs and entered Columbia's College of Physicians and Surgeons in the class of 1903. Two of his contemporaries in the class of '04 were his future colleague in bacteriology, Oswald Avery, who went on to prove that bare DNA was the stuff of genes, and Joseph Thomas, father of Lewis Thomas.

Zinsser cut his teeth in research while still in medical school, earning an MA in embryology and studying the effects of radium on bacterial survival. (I wonder about the relationship between radium research in those early days and Zinsser's leukemia.) His record earned him a prized internship at Roosevelt Hospital, then the primary teaching hospital of Columbia. Zinsser and the other house officers at Roosevelt lived in a small world of their own, sharing many of the hazing rituals of medical training and "walking out," as dating was then called, chiefly with the Roosevelt nurses, among them Grace Peck (see "Lewis Thomas"). "An interne who doesn't sooner or later fall in love with a nurse is usually a depraved fellow."[10]

The Roosevelt house officers in those days worked the wards, the clinics, and rode ambulances. William Carlos Williams was another intern at Roosevelt at the time, and Williams's account of his experiences in Hell's Kitchen[11] jibe with Zinsser's recollections. Zinsser's tale of answering a two-in-the-morning call to a tenement house could be the beginning of a William Carlos Wiliams poem:

> I found a man
> Who had been shot in the chest.
> He was lying diagonally
> across a small room
> lighted by a single gas burner.
>
> He was bleeding heavily into his clothes
> He also had a scalp wound
> Which bled profusely.
> My business was to get him to the hospital . . .
> I got him there alive . . . but he died.[12]

Zinsser picked up at Columbia what William Carlos Williams had acquired at Penn: the spirit of pragmatic humanism that flourished in those "aristocratic" American universities in the era of John Dewey. Williams's cry of "No ideas but in things" became an anthem of their generation. Its direct predecessor was James Russell Lowell's postbellum "Commemoration Ode." Zinsser would have seen a connection between Williams's plea for truth in things, and Lowell's ode to VERITAS:

> No lore of Greece or Rome
> No science peddling with the names of things,
> Or reading stars to find inglorious fates,
> Can lift our life with wings . . .
> But rather far that stern device,
> The sponsors chose that round thy cradle stood
> In the dim, unventured wood,
> The VERITAS that lurks beneath
> The letters unprolific sheath. . . .[13]

Medical schools stopped peddling with the names of things and started looking for VERITAS in the lab soon after the Flexner report of 1910. "Oh, Abraham Flexner!" Zinsser intoned. "We hail you the father—or, better the uncle—of modern medical education."[14] Flexner's exposé of the practitioner-dominated medical diploma mills that flourished at the turn of the century was the critical step in making American doctors members of a learned profession. Before then it was all words, words, words, and precious little experience: medical students sat on benches rather than working at them. After Flexner, American medical schools embarked on a century-long effort at empirical, laboratory-based medical instruction that became the envy of the world, "a phoenix rising." Zinsser would not have been pleased that nowadays—as in 1933 Germany—common sense is again becoming a victim of counterrevolution. In our day of the HMOs, of health care providers, insurance scams, and of "for-profit" hospital chains, he'd worry that the phoenix of our youngest science might return to the "ill-smelling ashes of a big business."[15]

When their days at Roosevelt Hospital were over, William Carlos Williams became a pediatrician in Rutherford, New Jersey; Joseph Thomas entered medical practice in Flushing; and Zinsser became a microbe hunter in academia. In 1915 Zinsser accompanied the American Red Cross Sanitary Commission to investigate a devastating outbreak of typhus in Siberia. After much trial and error, he eventually succeeded in isolating the European form of the microbe that caused typhus and worked hard at developing a vaccine against it. He also moved rapidly into the new science of immunology.[16]

His scientific gifts were not limited to microbe hunting, however. He also became a prolific medical writer and editor: his standard *Textbook of Bacteriology* went through many editions. Between 1928 and 1940 he regularly published poems in the journal first edited by James Russell Lowell and which Dr. Holmes had named the *Atlantic Monthly*: Dr. Zinsser later explained that his cultural life bridged "the period between Emerson and Longfellow to T. S. Eliot and James Joyce."[17]

After faculty positions at Stanford and Columbia, he was appointed to the chair at Harvard in 1923. Harvard at the time was a phoenix rising and its dean, David Linn Edsall, could truly report to his trustees that:

> There can be little doubt that the school has acquired the standing of being the best place in the country and perhaps anywhere for advanced training in research and for advanced training of teaching and research personnel."[18]

Until the Flexnerian revolution came to Boston, indeed from the time of its founding in 1782, Harvard's Medical School had functioned chiefly as a cozy nursery for Yankee practitioners. In the days of Dr. Holmes's deanship (1847–1853) the school offered little encouragement for laboratory research on the part of either faculty or students. In 1870, Dr. Holmes's clinical colleague, James Clark White, urged that rigorous standards of laboratory and bedside teaching—matching those of Paris, Berlin, or Vienna—be introduced in Boston: at the Harvard Medical School:

> When I find the young men of Europe flocking to our shores and crowding our native students from their seats and from the bedside, when the fees of our best lecturers are mostly paid in foreign coin, and when thousands of wealthy invalids from across the sea fill the waiting-rooms of our physicians, then I will confess that I am wrong, and that of the two systems of education ours is the best. Until then I shall seek in the spirit and working of their schools the secret of their success, the cause of our failings.[19]

Its provincial air had prompted the taunt that Harvard was the best medical school in Boston. But by the nineteen twenties, thanks to a new campus, a crop of young, full-time professors and an ebullient Dean Edsall, innovation was in the air. Rote learning had largely yielded to learning by doing, electives had sprouted, the experimental spirit had spread to the clinics, and Harvard was well on its way to becoming the best medical school in the country.

Edsall, a close friend and advisor to Abraham Flexner, assumed the deanship in 1918, took the job of medical education as a personal challenge, and set about recasting Harvard into a world-renowned academic medical center. Frugal Edsall told of snaring Zinsser from New York, bragging that "we got this professor for $2,000 less than we could have because of [his] personal income and because he wanted to come here."[20]

The "here" to which Edsall brought Zinsser in 1926 was that neoclassic campus north of the Charles. Its open piazza was framed by buildings well endowed with laboratory space for students and faculty alike. Edsall's medical acropolis attracted not only Zinsser, but also Otto Folin, Edwin Cohn, and John Edsall (the dean's son) in Biochemistry; Lawrence Henderson and Walter B. Cannon in Physiology; S. Burt Wolbach and Tracy Mallory in Pathology. Among the clinicians were William B. Castle, Soma Weiss, Herman Blumgart, and Maxwell Finland. His contemporaries also included future Harvard Nobelists: John F. Enders, Thomas Weller, George Minot, William P. Murphy, and George Hitchings. Zinsser was also in tune with Edsall's social views; despite major internal opposition the dean had appointed Alice Hamilton (1869–1970), a pioneer of industrial medicine, as the first woman assistant professor not only in the medical school, but in all of Harvard University.[21]

Zinsser's scientific career flourished at Harvard. His work on typhus carried him to Mexico and to China. In Mexico, he went after Rickettsia prowasekii that caused the disease and worked out innumerable approaches to a vaccine. He also detailed the epidemiology of a recurrent variant of typhus in European immigrants (Brill-Zinsser disease). Zinsser's work on typhus in Serbia, Mexico, and China spelled it out: lice require dirty humans, bad weather, and crowding—as in tents and barracks. That's why typhus is the stuff of war and tragedies and has—as he predicted—outlived Hans Zinsser. During World War II, typhus spread through North Africa, the Pacific Islands, and devastated Central Europe, where it was the second leading cause of death in the German concentration camps. On or about March 31, 1954, typhus killed Anne Frank in Bergen-Belsen only two weeks before the British Army came in to stamp out the lice.[22] American troops were protected by a vaccine based on the one developed by Zinsser and Castaneda and applied on a large scale. Although epidemic typhus declined at the end of World War II with the advent of DDT, R.prowasekii is making a comeback.

The largest recent outbreak since World War II was in Burundi in the mid-90s, where modern molecular techniques were used to show that a single outbreak of "jail fever" in Burundi sparked an extensive epidemic of louse-borne typhus in the refugee camps of Rwanda, Burundi, and Zaire—countries racked by ongoing civil war and genocide.[24] There was also a brisk outbreak in Russia in 1997. In Europe in the past and in Africa today, persons who have "recovered" from epidemic typhus in their youth suffer relapses of Brill-Zinsser disease and have become a reservoir of new, louse-borne epidemics.[23]

His work on typhus was not Zinsser's only contribution. He was also a pioneer in the study of autoimmunity, our allergy to self induced by microbes. He was drawn to its study by the leading cause of cardiac disease in the interbellum years: rheumatic fever. Zinsser had been studying allergy to the streptococcus for several years, and in 1925 published a seminal hypothesis that is now accepted wisdom. Entitled "Further Studies on Bacterial Allergy: Allergic Reactions to the Hemolytic Streptococcus" it argued that

> Failure to find the organisms themselves . . . suggested either a toxic or allergic pathogenesis. Such reasoning is especially applicable to the various forms of arthritis, in which it is at least logical to think of an allergic association.[25]

We now know that what we call "autoimmune" reactions to the microbe are indeed responsible, but still remain in the dark as to how the disease comes about. It remained for his student, Albert Coons to put us on the right track. Zinsser would have been pleased. In brief, Coons found that if one hooked a fluorescent molecule chemically to a purified antibody, one could then add this labeled antibody to samples of tissue and find the suspected antigen.[26] This powerful method, "immunofluorescence" was discovered by Albert Coons and Melvin Kaplan in the 1950s in order to test Zinsser's suggestion that rheumatic disease is due to what might be called friendly fire: our immune defenses against the microbe are launched against our own tissues because the strep and we share lookalike components. Our immune system is therefore tricked into treating the host as if it were an invader that required disposal. I'd argue that Zinsser laid the groundwork for much of what we know, or think we know, of the rheumatic diseases today.

Zinsser was as fine a teacher as he was a scientist. His infectious enthusiasm for pure science brought him the best and the brightest of students who eventually filled chairs of Bacteriology, Immunology, Medicine, and Public Health the world over. In turn, they've transmitted Zinsser's broad, humanistic concerns to their students. Lewis Thomas was one of those vectors, and I confess that I get twinges of what might be called a Brill-Zinsser recurrence of sentiment when I'm asked about a lifetime of teaching at a medical school:

> . . . as we grow wiser we learn that the relatively small fractions of our time which we spend with well-trained, intelligent young men are more of a privilege than an obligation. For these groups are highly selected and they force a teacher continually to renew the fundamental principles of the sciences from which his specialty takes off. So while we are, technically speaking, professors, we are actually older colleagues of our students, from whom we often learn as much as we teach them.[27]

Zinsser died of a hematologic malignancy. The last chapter of *As I Remember Him* forecasts in stoic detail the events of his terminal illness. He writes of himself in the third person:

> As his disease caught up with him, RS felt increasingly grateful for the fact that death was coming to him with due warning, and gradually. So many times in his active life he had been near sudden death by accident, violence, or acute disease. . . . But now he was thankful that he had time to compose his spirit, and to spend a last year in affectionate and actually merry association with those dear to him.[28]

Zinsser's legacy remains indelible in the two cultures. His well-written books remain in print and we live with his science. He taught us how we get typhus, created a successful vaccine against it, and told us how it can recur as Brill-Zinsser disease. He was the first to reckon that rheumatic diseases result from the friendly fire of our own armaments against microbes. But, his finest contribution was to warn us, from the field, from the podium, and in his writing, that "lice, ticks, mosquitoes and bedbugs will always lurk in the shadows when neglect, poverty, famine or war lets down the defenses."[29] The shadows remain in our age of Endarkenment.

Faith-based Alternative Medicine: Moses Applies for a Grant

AT THE HEIGHT OF HIS POPULARITY in 2002, President George W. Bush issued an executive order* that permitted faith-based social service organizations to apply for federal funding. With one stroke of the pen, the president had evened the playing field for major religions and minor universities. In the clinical realm, applications for funds were to be directed to the National Center for Complementary and Alternative Medicine (NCCAM). The NCCAM director, Stephen E. Straus, M.D., explained on its Web site that "Prayer and spirituality for the benefit of health are relied upon by many Americans. NCCAM seeks to develop strategies to bring the most rigorous and detailed scientific approaches to studying these and other CAM practices." NCCAM, founded in October 1998 by Congressional mandate, is staffed by an "estimated 80 FTEs" (full-time employee equivalents) and had a budget of $123,116,000 for FY 2005.

Applications were filed even as the president spoke; indeed one of these was leaked to the press. This grant submitted by the grandfather of faith-based social services, was reviewed by the NIH Alternative Diet Study Section.

---

* [*Author's note: the review is written in the format required from outside reviewers for federal grants from the National Institutes of Health.*]

Principal Investigator: Moses, son of Levi. Social Security #: 000-00-0001.

Organization: Mt. Sinai Department: Ambulatory Medicine

Title of Proposal: Dietary Control of Infective and Inflammatory Diseases

Reviewer: Higher Authority

Overall Summary: The principal investigator, Dr. Moses, is a veteran worker in the field who has been previously funded for epidemiological work. He has successfully published papers on the experimental induction of disease in Egypt: blood, frogs, gnats, flies, livestock death, boils, hail, locusts, darkness, and extermination of neonates (Ex. 1–40). A subset of the study population escaped to healthier climates. After preliminary data gathered at his Mt. Sinai laboratory, the principal investigator suggests new measures to prevent the spread of contagious disease. Quoting recent authorities from the literature, he proposes a series of strict dietary measures to prevent the outbreak of gastrointestinal and inflammatory diseases. The study population will be monitored by an angel who will guarantee constant follow-up by novel devices such as a cloud of dust by day (desert storm) and a pillar of fire by night. Basing many of his proposals on a series of experimental commandments, he will formulate public health measures for a chosen group of wandering nomads. He plans to study the effect of ten commandments over a forty years span of observation under desert conditions. His purpose is to test the maxim "Take two tablets and call me in the morning." No control populations are described.

Specific Critique—Specific Aim I (Dietary Control):
To make a difference between the unclean and the clean, and between the beast that may be eaten and the beast that may not be eaten. (Lev. 11:47)

Or if a soul touch any unclean thing, whether it be a carcase of an unclean beast, or a carcase of unclean cattle, or the carcase of unclean creeping things, and if it be hidden from him; he also shall be unclean and guilty. (Lev. 5:2)

The dietary methods proposed for the prevention of contagious disease seem reasonable, since it is well known that infective diarrheal disease can follow the ingestion of spoilt food. However, the specifics of the study parameters remain somewhat vague. For example, the PI proposes that one cannot eat the hare because *"he cheweth the cud, but divided not the hoof,"* and it is difficult to distinguish this creature (Lev. 11:4) from the camel who also *"cheweth the cud, but divided not the hoof"* (Lev. 11:6). Since neither camel nor hare are necessarily freer of parasites *(t. saginata)* than the cow, these strictures need further explanation. Moreover, it is difficult to understand why the PI argues for eating *"the locust after his kind and the beetle after his kind, and the grasshopper after his kind,"* but not the *"stork, the heron, the lapwing and the bat."* (Lev. 11:22–24). The inconsistencies extend to the aqueous terrain. *"Whatsoever has fins and scales"* are recommended but all that *"move in the waters without fin and scale are to be avoided."* Recent investigations of cholera and industrial pollutants suggest that teleosts accumulate as many microbes and toxins as shellfish. Nevertheless, the proposal may be a practical way to prevent the spread of viral hepatitis and other fecally transmitted pathogens. No serological testing is proposed and no prospective analyses are outlined. There is no statistical treatment of the insect population to be eaten.

Specific Aim II: (Inflammation)
When a man shall have in the skin of his flesh a rising, a scab, or a bright spot and it be in the skin of his flesh like the plague of leprosy; then he shall be brought unto Aaron the priest, or unto one of his sons the priests: (Lev. 13:2)

And the priest shall see him: and behold, if the rising be white in the skin, and it have turned the hair white, and there be quick raw flesh in the rising; (Lev. 13:10)

This aim is apparently a series of proposals to study the pathogenesis and transmission of cutaneous inflammation. In a well-described clinical setting, the principal investigator, Dr. Moses, outlines the diagnostic criteria for boils, furuncles, scabs, poxes, necroses and ulcers and argues for the diagnosis—by a physician (priest)—of the state of contagion of the lesion. This closely reasoned manual of desert dermatology is the strongest experimental point in his protocol.

And if the bright spot stay in his place, and spread not in the skin, but it be somewhat dark; it is a rising of the burning, and the priest shall pronounce him clean; for it is an inflammation of the burning. (Lev. 13:28)

This important verse is not only the first description of inflammation, but also suggests that inflammation is a useful aspect of host defense. It is clear that an understanding of how boils spread can help prevent impetigo and the travail of nomad tribes. It is less clear how the dietary practices described in Aim I relate to the dermatological practices described in Aim II. Nor does monotheism seem the only solution for furunculosis. Again, no statistical treatment is presented and the PI seems content with anecdotal evidence.

Specific Aim III: (Faith-based Social Program)

1. *Thou shalt have no other gods before me.*
2. *Thou shalt not make unto thee any graven image . . .*
3. *Thou shalt not take the name of the LORD thy God in vain;*
4. *Remember the sabbath day, to keep it holy . . .*
5. *Honour thy father and thy mother . . .*
6. *Thou shalt not kill.*
7. *Thou shalt not commit adultery.*
8. *Thou shalt not steal.*
9. *Thou shalt not bear false witness against thy neighbour.*
10. *Thou shalt not covet thy neighbour's house, thou shalt not covet thy neighbour's wife, nor his manservant, nor his maidservant, nor his ox, nor his ass, nor any thing that is thy neighbour's.* (Ex. 20:3–17)

This specific aim appears to fulfill the charge of the National Center For Complementary and Alternative Medicine (NCCAM) which states that "associated belief systems" may be required for these practices to succeed. In this reviewer's opinion, these commandments constitute such a belief system and soundly conform to the president's order. The first five appear to be concerned mainly with reaffirming the authority of the principal investigator and his higher authority at Mt. Sinai. The day of rest is a novel element but fails to provide for professional football. The last four commandments seem entirely impractical, since it is clear that in Washington, D.C., they have been honored more in the breach than in observance. Politicians of all major political parties have not failed to steal, bear false witness against their neighbors—and certainly not to commit adultery. The temptation to covet a neighbor's house, wife, manservant, maidservant, ox, or ass is the basis of commercial television. However, it is difficult to see how these commandments relate to the primary purpose of disease prevention. Moreover, the PI leaves unclear how his belief system can be extrapolated to other tribes, nor to days other than the Sabbath. No statistical treatment is offered in this section.

Investigator: The principal investigator, Dr. Moses, has had a productive early track record (e.g., Ex. 1–40). In addition to his well-regarded work on experimental induction of disease (see above) he has turned water into blood, made brick without straw, parted the Red Sea, fed his people with manna and quail, brought water from the rock, and seen God in a burning bush. Unfortunately, he hasn't published very much recently. Is he burning out? The PI allows that it will take thirty-nine more years for his people to reach the Promised Land and has described the many personnel problems created by his absence for forty days and nights at Mt. Sinai. During that time the study population appears to have been diverted by a golden calf. It is unlikely, therefore, that the principal investigator's rigid code of behavior will maintain order and prevent disease for a span of forty years. His own data show that lightning and thunder from Mt. Sinai did not suffice to maintain order for forty days.

The principal investigator's chief assistant, Aaron, has a poor publication record. So far he has simply described the transformation of wooden staffs into serpents, a trick anticipated in the Egyptian literature. Aaron has never been independently funded and is the recipient of a few paltry grants from pharmaceutical manufacturers. He is no Moses. Should something happen to the principal investigator, this ambitious forty-year study may well falter in the desert.

Overall Critique: The proposal seems to confuse the dietary management of disease with the establishment of monotheistic authority. The PI summarizes his plan: *Ye shall not eat any thing with the blood; neither shall ye use enchantment . . .* (Lev. 19:26), a proscription reinforced by the argument: *Regard not them that have familiar spirits neither seek after wizards, to be defiled by them.* The principal investigator here argues against supernatural tricks proposed by wizards other than himself (*viz*, the burning bush). That doesn't seem, shall we say, kosher. The overall proposal has some merit with respect to pork and infectious disease but fails to establish criteria for evaluating the validity of one faith-based dietary regimen (e.g., vegan, macrobiotic) over another. Moreover, the principal investigator has made no provisions for minority groups, and women's health issues are not addressed.

Human Subject Considerations: The requirements for 600,000 men, women, and children in previous studies (Ex. 1–40) are somewhat excessive. The 3,000 Israelites slain by Aaron and the priests for worshipping the calf of gold do not meet the criteria set for violations of commandment #2. Informed consent forms have not been enclosed.

Budget: One questions the allocation of fifteen cubits of hangings for the temple and the high cost of onyx and jasper for the sanctuary. It is not clear that the vessels for the altar with the "cloth of blue" are necessary for the second year of the grant and the covering of badger's skin for the curtains seems excessive (Ex: 40:39). Moreover, such items as the gold candlesticks, golden spoons, silver chargers, and snuff dishes (Num. 4:9–25) could well be deleted.

Research Environment: Hot, sandy, windy, with wet periods infrequent. Scorpions and serpents. Animal quarters for sacrificial beasts are inadequate. The pillar of salt is unlikely to work as a means of electrolyte replacement.

Overall Rating: The reviewer recommends approval with an average degree of enthusiasm.

Priority Score: 2.8

[*Author's note: another way of saying "Your call."*]

Reducing the Genome*

Since the Renaissance there have been no complete breaks in the continuity of the sciences comparable to those sterile periods which have, from time to time, interrupted the progress of the arts.

—HANS ZINSSER, *As I Remember Him*, 1940

IN THE LAST HALF of the last century science explored the edge of the universe, the deepest vents in the sea, and cloned sentient life in Scotland. Meanwhile, progress in the arts—to echo Zinsser—was interrupted by the age of Jacques Derrida and Jeff Koons. Looking back, I wonder how many would now agree that a symposium called "Our Country, Our Culture" organized by *Partisan Review* was the critical intellectual event of 1952. A better contender might be the three-dimensional model of DNA. The double helix was the brainchild of a raw, Midwestern postdoc and a renegade physicist in the capital of crystallography, Cambridge. After Maurice Wilkins had offered them a long peek at Rosalind Franklin's X-ray data in 1952, James Watson and Francis Crick came up with a model for DNA that correctly placed the phosphate groups on the outside and the four bases (A and T and G and C) neatly paired on the inside of the helix. The structure appeared in *Nature* of April 26, 1953,[2] just a year after the *Partisan Review* symposium.

I'd suggest that in the fifty years since, our country and our culture have been influenced at least as much by the description of the double helix and its practical consequences as by the intellectual concerns addressed in *Partisan Review*. Two recent extensions of DNA science have jarred the temper of our time. The first was the reproductive cloning of sheep Dolly by Ian Wilmut in Scotland (1997)[3] and the second was the deciphering of the complete human genome by Celera Genomics[4] and a public consortium[5] (2001). Our country and our culture have responded by resurrecting what John William Draper called *The Conflict Between Religion and Science* (1874).[6]

---

* On the occasion of the fiftieth anniversary of the 1952 Symposium "Our Country, Our Culture" held, in Boston, under the same title, by *Partisan Review*, June 2002.

The Genome Race

The imminent success of the human genome project was guaranteed at a Washington press conference in June of 2000. From the podium, President William Jefferson Clinton told the world that reading the human genome was "like opening up the book of life." On his right stood Craig Venter, head of Celera, the private company that had set the pace of the quest. To the president's left was Francis Collins, leader of the NIH—or public—arm of the Human Genome Project. The two scientists had arrived at that podium by different paths. Collins is a self-confessed, born-again Christian, who once told science writer George Liles that

> You've got to accept who Christ was and what He said, or reject the whole thing. . . . I do think that the historical record of Christ's life on earth and his Resurrection is a very powerful one.[7]

Sentiments of this sort seem to have made Francis Collins palatable to those in Congress who might not otherwise favor governmental support for genetic research. One notes that fans of Holy Writ have recently pushed reproductive cloning offshore and placed a moratorium on therapeutic cloning.

Craig Venter, on the other hand, is an entrepreneurial, no-nonsense, ex-Navy corpsman, who insists that "the human genome is not the book of life, it is not the blueprint of humanity, it is not the language of God, and it is certainly not the parts list of humanity."[8] He tends to agree with British scientist Sidney Brenner that "President Clinton described the human genome as the language with which God created man. Perhaps now we can view the Bible as the language with which man created God."[9]

Francis Collins and his ilk notwithstanding, I'd argue that the vis a tergo behind the ascent of science in our time is what Jacques Loeb called *The Mechanistic Conception of Life* (1912), or reductionism redux, if you will.[10] It's the notion that life—if not life's value—can be explained by the laws of chemistry and physics, and that, à la William Carlos Williams, there are "No ideas but in things." Loeb and the mechanists believed that revolutions in science have enlarged the place of fact in a world of value. Many of us agree with the mechanists that Western, experimental science is by definition, uni- and not multicultural: scientists from many cultures produce the single culture of science. In that empire of fact, each can claim *"Civis Scientium sum."*

I've argued that we've made such dramatic progress in our understanding of living things that over the last half century we've brought about a

"Biological Revolution" (1976).[11] That revolution, it now turns out, has been directed as much by entrepreneurial interest as by academic curiosity; the invisible hand of Adam Smith is firmly on the tiller. Craig Venter's company was launched with the motto "Speed Matters," and Celera made speed matter, indeed. The Human Genome Project was formally launched in 1987 by the Department of Energy with support for this effort from the National Research Council/National Academy of Sciences. The NRC/NAS estimated early in 1988 that deciphering the genome would cost three billion dollars, and be finished as early as 2003. The public consortium, headed by James Watson, soon included labs of the DOE, the National Institutes of Health and international brigades from France, Britain, and Japan. By 1993, the spiritually correct Francis Collins took over leadership of the public effort and promised results by 2005.

Craig Venter, on the other hand, had started out at the NIH, but soon grew impatient with the pace of the affair. After working out novel methods for expressing and sequencing genes ("expressed sequence tags" and "the whole genome shotgun method"), Venter founded Celera Genomics in 1998 vowing to unravel the human genome better and faster. He estimated that it would cost Celera about 250 million dollars and that he'd be finished in 2001. He raised the money, harnessed new sequencing machinery, and the race was on. The consortium struck back in a hasty effort to recoup. By August of 2000, both groups announced that the map of the human genome was pretty much finished and in January 2001, joint, if slightly disparate maps were published. Bottom line: public effort: 14 years, two billion dollars; private effort: 2 1/2 years, 250 million dollars.[12]

The Encyclopedia Race

Was there anything like this before in human history: a national competition, in which a small group of private entrepreneurs beat out a huge public consortium headed by perhaps the finest scientist of his day? Actually, there is a precedent, and it's the story of the *Encyclopédie*. In 1675, Colbert asked the *Académie des Sciences* to prepare an illustrated catalogue of the arts and metiers of the nation. The best scientist of his day, René Antoine Ferchault de Réaumur was charged with the official task: Réaumur had invented the thermometer, was a dazzling mathematician, and stood astride the gates of physical sciences in France. Three score years passed without a publication—or a rival. The competition began when in 1747, Diderot and the Abbé Gua de Malves signed on with a commercial publisher for a new encyclopedia based on *Chamber's Encyclopedia* of London. Diderot had cut his teeth on the unprofitable *Dictionnaire de médecine* (1746); he now assembled his fellow *philosophes* as *La Société de Gens de Lettres* to prepare and distribute the more ambitious *Encyclopédie* to subscribers, among them the gentry of Enlightenment Europe.[13] By 1751, the first volume of Diderot's *Encyclopédie, ou Dictionnaire raisonné des sciences, des arts et des métiers* was issued, with a promise to subscribers that an appendix of plates would soon be forthcoming.

By 1759 not one fascicle of the Academy's plates had been published, although some had been engraved and many more drawn. That year, Diderot and crew started their own efforts at engraving, having looked over some of the Academy's plates as possible models. In the grand tradition of French irascibility, one of Diderot's disgruntled workmen accused Diderot of taking 77 plates from Réaumur's advance proofs, and the Academy sent a commission to scour Diderot's premises. They found at least 40 of Réaumur's unpublished plates, and made Diderot agree to show the Academy each of his own new plates before publication to make sure that none were filched from the Réaumur set. None, indeed, were, and the two sides published the first volume of plates simultaneously in the winter of 1761.[14] Bottom line, public effort: 87 years for plates and text; private effort: 10 years for the plates, 4 years for the text. Speed mattered in the Enlightenment as the *Encyclopédie* became the *machine de guerre* of reductionist science.

Nowadays in our country and our culture, Enlightenment science faces opposition from the political right and left; both factions share a faith-based fear of science. The admirers of nature, eco-sentimentalists, like Bill McKibben with his *The End of Nature* are joined by those McKibben called "Unlikely Allies Against Cloning":

> The changes engendered by genetic engineering . . . could force us to reconsider liberalism's faith in the onward march of science. This message was made clear last week when a broad coalition of environmentalists, feminists, and other progressives released an open letter urging the Senate to ban reproductive cloning and to place a moratorium on therapeutic cloning. . . . At one such rally of environmentalists against sports utility vehicles last spring, a minister held a sign that read "What Would Jesus Drive?"[15]

John William Draper

Faith-based fear of science has a long history. As late as 1854, doctors in New York State were not permitted to dissect the human body in medical schools. A statute called "The Bone Bill" passed the legislature by a whisker, only after bitter controversy between defenders of "natural law" and proponents of medical sciences. The doctors pleaded that Vesalius's great dissection atlas *De Humanis Corporis Fabrica* had been used as a manual by medical students in more enlightened parts of the world since 1543. But in 1854, *Harper's Weekly* argued the antidissection cause in phrases that sound very much like the Bush administration's stand against therapeutic cloning:

> Science may prove ever so clearly that there is nothing there but carbon, oxygen, and lime . . . but all this can never eradicate the sentiment we are considering, and that is too deeply in our laws of thinking, our laws of speech, our most interior moral and religious emotions.[16]

Interior moral and religious emotions yielded to medical science when antebellum Manhattan finally caught up with Renaissance Brussels. The driving spirit behind the Bone Bill was John William Draper, founder and first president of New York University School of Medicine. Draper sounded the tocsin for the progressive side:

> The history of science is not a mere record of isolated discoveries. It is a narrative of the conflict of two contending powers—the expansive force of the human intellect on one side and the compression arising from traditional faith and human interest on the other. As large a number of persons now live to seventy years as lived to forty, three hundred years ago.[17]

I would note that mean longevity in the United States in 1840 was about 40 years. By the 1980s, it was approaching 80 years.[18] That doubling of human longevity, as Draper would have been the first to argue, can hardly be credited to major increments in "our most interior moral and religious emotions."

Draper not only wrote a book that became the *machine de guerre* of free thought for the better part of a century, but as Professor of Chemistry at NYU, shared with Samuel F. B. Morse, Professor of Fine Arts at NYU, the honor of producing the first daguerreotype portraits by an American. In 1840, the two entrepreneurial professors opened up a commercial photography studio at the top of the University building in Washington Square; they also taught photography to the likes of young Matthew Brady. They collaborated successfully on another project at NYU in the 1840s: they reduced words to the dots and dashes of the Morse code in their Washington Square lab. Tinkering with an early version of his telegraph, Morse called on Draper to help design cables that might transmit electromagnetic signals over distances longer than that of their studio perimeter.

Draper framed and tested the hypothesis that the conducting power of an electric wire varied directly with its diameter; an equation that made commercial telegraphy possible.[19] When in 1844 Samuel F .B. Morse transmitted the message "What hath God wrought?" from Washington to Baltimore he might just as readily have asked "What hath Draper wrought?"

During the American Civil War, of which he wrote the first full history, Draper was one of the commissioners appointed to inspect hospitals after the battles of Antietam and Gettysburg. Matthew Brady was his photographer. Draper's prescient *Textbook of Physiology* recognized that our sensations of color, vision, and time were relative, that subjective impressions frame the facts of nature, as they might in a dream.

> One of the most extraordinary phenomena presented in the dreaming state, is the instantaneous manner in which a long series of events may be offered to the mind, the exciting cause being truly of only a momentary duration. Some sudden noise arouses us, and in the act of waking, a long drama connected with that noise appears before us; or in like manner . . . we are disturbed perhaps by a flash of lightning, and with that flash occurs a dream which seems to us to occupy a space of hours or even days, so many are the incidents with which it is filled.[20]

In 1847 Draper had published his "Production of Light by Heat," an important and early contribution to spectrum analysis, indeed he was the first to photograph the diffraction spectrum. He was also the first to take a photograph of the moon, and with his son, to prepare the first photomicrographs of tissues. Draper was also a student of older cultures, and delivered a series of three archeological lectures on "Egypt and Egyptian Antiquities" before the New-York Historical Society in December of 1864. He concluded, from a perusal of those antiquities, that "public education and intercommunication" were the only sure bases of a stable society."[21]

Draper was proud to have been made a foreign member of Galileo's secular Accademia Nazionale dei Lincei in Rome. Ever the polymath, he became the featured lecturer at that historic British Academy debate at Oxford in 1860 between evolutionists and their religious foes. Draper was invited as the major American champion of his fellow Linceians, Charles Darwin and Thomas Huxley; meanwhile the doyen of American biology, Louis Agassiz, was still flying the banner of creationism at Harvard[23] (see "Swift-boating Darwin").

Looking back at that Victorian battlefield, Owen Chadwick, a Cambridge historian of religion, dismissed Draper as a shallow village atheist:

> Draper's books contain the paean of praise to science, a hymn, its mighty achievements, among them the telegraph, telescopes, balloon, diving bells, thermometer, barometer, medicines, railway, air pump batteries, magnets, photographs, maps, rifles, and warships. . . . Draper never stopped to ask himself why anyone who invented a camera or possessed a barometer might be led to think his faith in the God of Christianity shaky.[23]

I'm convinced that while the twentieth century mind may not be secularized by telescopes, balloons, diving bells, and thermometers—or surfing the internet for that matter—a glance at the bills of mortality might lead aging academics to show at least some respect for reductionist science. Recall that in the fifty years since the *Partisan Review* conference was held, the longevity of Americans has increased by about a decade and a half. But the rate of that increase since 1952 isn't as steep as the decline in the rate of deaths. The obvious conclusion is that our country is aging. One might also conclude that we're not living longer because of Bishop Wilberforce's belief system, but because of what medical science has learned over the years thanks to contributions like Draper's Bone bill, his Physiology text, and his photomicroscopy. Woody Allen might have had Draper in mind with his wisecrack from "Deconstructing Henry": "Science is good. Given a choice between air-conditioning and the Pope, I'll take air-conditioning."

FISH and chips

The Lynxes and Fellows of the Royal Society started the materialist ball rolling with their Horatian mottos. They also laid the groundwork for a mechanistic conception of life fully articulated by Jacques Loeb over two hundred years later. Moreover, Hooke and the Royals appreciated that an entrepreneurial spirit was essential to the tempo of accelerated discovery:

> . . . the Real esteem that the more serious part of men have of this Society, is, that Several Merchants, men who act in earnest (whose Object is meum-teum [free enterprise], that great Rudder of humane affairs) have adventur'd considerable sums of Money, to put in practice what some of our Members have contrived, and have continued stedfast in their good opinions of such Indeavours, when not one of a hundred of the vulgar have believed their undertakings feasable.
>
> And it is also fit to be added, that they have one advantage peculiar to themselves, that very many of their number are men of Converse and Traffick; which is a good Omen, that their attempts will bring Science from words to action, seeing the men of Business have had so great a share in their first foundation.[24]

In our era we bring words to action by means of imaging techniques that we owe to the Royal Society and to the men of Converse and Traffick who put in practice "what some of our Members have contrived." The Royals' *Nullius in Verba* led the way from Newton's *Opticks* to Hooke's *Micrographia* to Helmholtz's lenses, to Carl Zeiss, Inc., to Bausch & Lomb, and to Hewlett Packard. So now we have an optics that shows how chromosomes are dispersed and rearranged: "fluorescence in situ hybridization," or FISH, for short. This method permits us to show genes flashing in orange and green, like sunlight on a Monet haystack. In like fashion, molecular biologists have fulfilled the dream of another Royal, Erasmus Darwin, Charles's grandfather. They have rapidly exploited the promise of the genome project by dropping genes on tiny chips. By this means, we learn which genes are up- or down-regulated in any area of any tissue by means of microchips on which thousands of genes are displayed. One can spot patterns of aging, infection, or cancer as tiny rows of refracted fluorescence shimmer on the chip. They look like summer on the Grande Jatte.

But the revolution of genes is not fueled by FISH and chips alone. We can now make tissues in a dish and move DNA around at will. It's a worrying notion that has troubled folks mightier than biologists. Pope John Paul II has warned a crowd of at least two million Poles that "mankind was going dangerously astray by letting scientific advances and cultural liberalism eclipse God's will."[25] That great and holy man was troubled that modern "genetic manipulation" might displace "the Creator's right to interfere in the mystery of human life." As I read the pontiff's warning, I began to worry.

Almost forty years ago in Cambridge, Alec Bangham and I had distilled lipids in a glass vessel to form liposomes, artificial cell membranes that have now been used in the clinic to deliver drugs and enzymes. Indeed, by encapsulating DNA into liposomes, one can introduce genes into cells or form self-replicating DNA in a dish. The process is called lipofection and has become vastly popular in labs worldwide.[26] Distressed by the pontiff's warning that we were interfering with the mystery of human life, I turned to older authority. I was reassured that our liposomal legerdemain had been foreseen by Denis Diderot in his *d'Alembert's Dream.*

> You can distill life in a closed vessel. Eat, digest, distill in a closed vessel, and you have the whole art of making a man. Anyone who wants to describe to the Académie the production of man or animal will need to make use of nothing but physical agencies.[27]

Diderot was not the only Enlightenment scholar who correctly predicted the facts of biology in epigrammatic form. In year Ten of the revolution, Pierre-Jean-George Cabanis suggested "*le cerveau secrète le pensée comme le fois secrète la bile.*" ("The brain secretes thoughts like the liver secretes bile.")[28] Cabanis's formulation was right on target. Scientists studying the "mirror neuron system" have found discrete areas of the brain that clearly emit metabolic signals during mimetic behavior: thought *is* secreted.[29] Finally, Eric Kandel was awarded the Nobel Prize in 2000 for showing that the release of neurotransmitters from synaptic vesicles at nerve terminals is in fact guided by chemical signals of the sort that control secretion by liver cells.[30] The brain, literally, secretes thought.

The history of Western science ever since Galileo and the Linceians (*eppur si muove!* and still it moves) has been enriched by philosophers, chemists, and physicians generally out of favor with the legions of God. By 1855, Jakob Moleschott had established the mechanist's motto: "No phosphorous, no thought."[31] We now know that the slogan was smack on. Paul Greengard, who shared the prize with Eric Kandel, showed that the state of phosphorylation of a molecule known as DAPP 32 regulates our moods, our feelings and our responses to drugs.[32] These days, "No phosphorous, no thought" sounds like a *reductio ad profundum.*

Hippolyte Taine: The Reduction to Images

Reductionism is the *sine qua non* of modern evolutionary science. It is no accident that its antonym, "holism," was coined de novo in an anti-Darwinian tract called *"Holism and Evolution"*[33] written in 1926 by the Boer general, Jan Christian Smuts, a defender of Owen Chadwick's brand of theology. Reductionism can be traced to Hippolyte Adolphe Taine, historian, literary critic—and another Linceian—who introduced the term reductionism in his 1872 book *de l'Intelligence*:

> Since our ideas may be reduced to images, their laws may be reduced to laws of images. Images, then, are what we must study.[34]

Taine, whose literary sensibility was vast (he wrote a multivolume *History of English Literature*) would have been in tune with William Carlos Williams's reductionist plea "No ideas but in things." Taine's laws of images are in direct line of descent from the refraction of light in Newton's *Opticks*, Helmholtz's color theories, Draper's spectral photographs, and Morse's reduction of words to dots and dashes on the wire.

We owe to them the molecular biologist's FISH and microchips: the optical, digital, and kinetic techniques we use to ferret out the facts of science. Taine relied on molecules before molecular biology was born:

> By means of such evidence, and of the recent discoveries of physicists and physiologists, we have attempted . . . to conceive the connection of molecular nervous changes with thought.[35]

Finally, it is probably no accident either that the terms reductionism (Taine) and impressionism (Louis Leroy, Jules-Antoine Castagnary) arose at the same time. Here is the art critic:

> If one wishes to characterize and explain [the paintings of Monet et al.] with a single word, then one would have to coin the word impressionists. They are impressionists in that they do not render a landscape, but the sensation produced by the landscape. The word itself has passed into their language: in the catalogue the Sunrise by Monet is called not landscape, but impression. Thus they take leave of reality and enter the realms of idealism.[36]

Taine's shimmering prose in the following passage is the mirror image of Monet's first *"Impression: soleil levant"* of 1872. Both are products of an upbeat, experimental attempt to reduce a mass of blooming, buzzing confusion to beautiful bytes of sense or pigment:

> Yesterday about five, I was on the quay by the Arsenal, watching in front of me, across the Seine, the sky reddened by the setting sun. Fleecy clouds rose in the form of a half dome, and bent over the trees of the Jardin des Plantes. The whole of this vault seemed incrusted with scales of copper; countless indentations, some almost burning, some nearly black, extended, in rows of strange metallic lustre, up to the highest part of the sky, while, all below, a long bronze-colored band, extending along the horizon, was streaked and cut by the black fringe of branches. Here and there, rose-colored gleams of light rested on the pavements; the river shone softly through a rising mist; I could see barges floating with the stream, and two or three teams of horses on the bank, while towards the east, the slanting beams of crane stood out against the gray sky. In half an hour, all this had died out; there was but one patch of clear sky behind the Pantheon; reddish-colored smoke was wreathing about in the dying purple of the evening, and the vague colors intermingled.[37]

Taine reduced the notion of time passing to these visual images, as Draper reduced moonlight to a glass negative, and Morse our speech to an electric code. But reductionism has not robbed ideas, images, or speech of their power to move, inform, and enlighten us. Indeed, the reduction of ideas to image and of speech to bytes has permitted us to experience the shock of the new, be it Monet or the map of our genome.

A Nobel Error

In December of 2001, when the Nobel Foundation celebrated its centenary, most of the living Nobel Laureates in Physiology or Medicine showed up to revisit their grandest moment and to take part in a gala celebration. The Nobelists included a good number of American physicians (Barry Blumberg, Alfred Gilman, Joseph Goldstein, Michael Brown, Eric Kandel, among others). The occasion was celebrated by lectures, symposia, concerts and the grandest of all banquets in the splendid Town Hall of Stockholm. I was lucky enough to have been invited to the centenary and luckier still to have been placed at the banquet table with the senior American Nobelist there, Tom Weller of Harvard. Weller, the most modest and genial of physician/scientists, had shared the prize in 1954 with John Enders and Fred Robbins "for their discovery of the ability of poliomyelitis viruses to grow in cultures of various types of tissue."[1] Their work made possible the development of the Salk and Sabin vaccines against polio. Over wines of not inconsiderable vintage, and loaded with resveratrol, Weller reminded his dinner companions that the first time American doctors had been feted in this hall, it was for a dazzling cure based on a fundamental error. In 1934, two Harvard clinicians, George R. Minot and William Murphy, joined George C. Whipple, a Rochester pathologist, on the podium in Stockholm. Based on the wrong animal model, they had found a cure for pernicious anemia.[2]

George Richards Minot was "a Yankee of the Yankees," culturally an aristocrat, politically a conservative, but in behavior a democrat. Minot was born in 1885 to a family of Boston Brahmins; his father, James Jackson Minot was named after a physician and became a physician himself; Minot's mother was Elizabeth Whitney. An early naturalist, Minot had published two articles on butterflies before entering Harvard; his AB degree in 1908 was followed by the M.D. in 1912. After house-staff training at Massachusetts General Hospital and Johns Hopkins Hospital, he returned to Harvard's Peter Bent Brigham in 1915 and by 1928 Minot succeeded the legendary Francis Weld Peabody as chief of the Harvard Medical Unit at Boston City Hospital. He also maintained an active private practice oriented to hematology.[3]

By way of contrast, William Parry Murphy, with whom Minot shared the Nobel prize, was a Westerner of decidedly nonpatrician stock. Born in 1892, at Stoughton, Wisconsin, the son of a congregational minister with various pastorates in Wisconsin and Oregon, young Murphy was educated in the public schools. He took his AB degree at the University of Oregon in 1914 and for the next two years taught physics and mathematics at several high schools in Oregon. After one year at the University of Oregon Medical School he won a scholarship to the Harvard Medical School from which he graduated in 1922. Two years as a house officer at the Rhode Island Hospital were followed by an eighteen-month residency at the Peter Bent Brigham Hospital.[4] It was there, in early 1925, that Minot made Murphy an offer he could not refuse.

Minot, so the story goes, was accustomed to picking young physicians of the Peter Bent Brigham as associates to run his office practice, which consisted in good part of patients with diseases of the blood and with homes on Beacon Hill. As senior resident, Murphy was next in line for this plum job, but it was customary for the young men to earn their credentials by publishing one or more papers before they started. So Minot suggested to Murphy—who was by no means committed to clinical investigation—that he had better find some project or other in hematology before joining his practice. Murphy took to the journals and found a recent report by George Whipple and Frieda Robscheit-Robbins[5] that dogs made anemic by repeated bleeding could be restored to health by feeding them huge quantities of uncooked liver. If it worked in dogs, why not in humans? The first anemic patients to whom Murphy fed slightly cooked liver were patients with pernicious anemia, a disease that not only affected their marrow, but also perturbed their mental functions. One of these patients was an obstreperous, fretful, cantankerous old woman, whom Murphy cajoled into taking her daily ration of half a pound of liver only after a mighty contest of wills. To Murphy's everlasting surprise, not only did her red blood cells respond to a week or so of this cumbersome regimen, but she was also relieved of her mental symptoms and soon reverted to her agreeable self. The factor in liver was therefore something that worked not only in the marrow but also elsewhere in the body.

By May of 1926, Minot and Murphy had treated each of forty-five patients with half a pound of liver a day, and a report detailing their miraculous responses was presented to the Association of American Physicians in Atlantic City.[6] But the whole enterprise was based on a false premise, an error of interpretation. Whipple's dogs by no means suffered from pernicious anemia: it was pure iron deficiency due to the repeated bleedings, and their response was no doubt due to the iron present in the massive doses of liver the dogs were given. It may have been an error, but it led straight to the Nobel Prize.

Soon, Minot persuaded a young biochemist, E. J. Cohn, to make a liver extract rich in the anti–pernicious anemia factor. The extract proved to be a powerful stimulus to red cell production by the marrow[7] and it was Frank Peabody who found out how to measure the success of liver extracts. By careful analysis of bone marrow aspirates (then a new diagnostic tool) Peabody had discovered that pernicious anemia was due to failure of red cell maturation rather than destruction.[8] The bone marrow was filled with immature red cells but there were precious few red cells entering the blood stream. The nursery was full, but the kindergarten was empty. Peabody concluded that the disease was probably due to a nutritional deficiency, and not to some toxin or infection. There must be a factor in liver that "promotes the development and differentiation of mature red blood cells."[19] It followed that if one wanted to isolate an anti–pernicious anemia factor, one had to count the number of new red cells (reticulocytes) in the circulation: one had to know how many kids entered kindergarten.

Peabody's contribution was critical to the extension of Minot and Murphy's discovery and the eventual isolation of vitamin B_{12} as the extrinsic anti-pernicious anemia factor.[10] Edwin Cohn's semipurified extracts constituted the primary treatment for pernicious anemia until 1948, when extrinsic factor was isolated by chemists at Merck and Glaxo and named vitamin B_{12}. This red, crystalline substance was found to be a naturally occurring, abundant material in the flesh or milk of ruminant animals. Soon pernicious anemia could be cured by monthly injections of vitamin B_{12} averaging only one thousandth of a milligram and costing less than penny a day.[11] It was among the first "miracle drugs" that could cure a hitherto fatal disease. And it was discovered because of an error with some dogs fed liver.

There's more to the Minot story, as Weller recalled, another miracle drug had played a role. Not only was Minot a careful clinician, a cool intelligence, and a formidable presence; he was also gravely ill when he made his great discovery. A strapping six-footer, Minot had developed severe diabetes between 1921 and 1922. Since the only treatment in those preinsulin days was a very restricted low-sugar diet, his weight had dropped to 120 pounds and he was in bad shape thanks to disease or therapy, or both. As luck would have it, by 1922 Banting and Best working in John Macleod's lab in Toronto, had prepared the first useful batches of insulin. Early in 1923 Minot's personal physician, Eliot P. Joslin, obtained some of the precious material and Minot became one of the first patients treated with insulin. By trial and error, Minot adjusted his daily dietary schedule to the demands of repeated insulin shots, eventually regaining some of his earlier vigor. Years later, William B. Castle, his successor at the Thorndike, suggested that "it was not without significance for the rigorous subsequent trial of a special diet [for pernicious anemia] that Minot had become, perforce, compulsively concerned with details of his own [for diabetes]."[10] In short, had insulin not been isolated, B_{12} might have been a distant dream. Banting and Macleod's Nobel Prize in 1932 was followed by Minot and Murphy's Nobel Prize of 1933. It was a time of medical miracles in North America.

Murphy remained a practicing physician on the staff of the Peter Bent Brigham Hospital for the remainder of his life. And, for decades, Harvard medical students were astonished to learn that the gentle, soft-spoken hematologist who taught them how to listen to their patients had once been honored in Stockholm. Weller recalled that Murphy was "a sweetie, he always thought the best of everyone." Yet Murphy never attained a professorship at Harvard; from 1928 until 1935 he remained an Instructor, the lowest rank above house officer, and from 1935 until 1938—well after the Prize—he was made an Associate in Medicine. From 1948 until 1958 he became a Lecturer in Medicine, in 1958 a Senior Associate, and subsequently an Emeritus Lecturer. Not much of a CV compared to Minot's.

It was a different life, but, from reading his clinical notes one can get a sense of what Murphy was doing all those years. One can guess why Murphy, the minister's son, might have been satisfied simply to understand how humans get sick rather than to grab the mere stuff of rank. Here is an excerpt from one of Murphy's charts, written on January 14, 1926; it's not the hematocrit that concerns him, but the story of his patient:

> SOCIAL HISTORY: The patient is a single machinist who occupies one room in a boarding house. He does not pay rent while in the hospital. He has an aged mother who is apparently only partly dependent upon him. During his disease other folks will take care of her. The patient has no home or place suited for convalescence. He is of German descent, was born in Nova Scotia. He came to this country at the age of one and has no language difficulties. He is fairly well educated and is intelligent enough to follow dietary instructions and can pay for the same. No emotional factors in the patient's home at the present time.[11]

The tale is neat, the listener attentive, the prose is clean. The sound is that of William Carlos Williams, not of the *New England Journal of Medicine*: Murphy, not Minot. But after all, or so the story goes, Murphy was the one who made that Nobel error in the first place.

The Mother of Us All: Boston City and the Thorndike

> Ye worthy, honored, philanthropic few,
> The muse shall weave her brightest wreaths for you.
> Who in Humanity's bland cause unite,
> Nor heed the shaft by interest aimed or spite;
> Like the great Pattern of Benevolence,
> Hygea's blessing to the poor dispense;
> And though opposed by folly's servile brood,
> ENJOY THE LUXURY OF DOING GOOD
>
> —C. Caustic (nom de plume), *Terrible Tractoration: A Poetical Petition Against Galvanizing Trumpery. . . .* 1804[1]

DOCTORS COME OF PROFESSIONAL AGE in their house-staff years: habits are rehearsed, skills honed, friendships set for life. Life as a resident in a teaching hospital combines aspects of an army recruit's basic training with a lawyer's clerkship in chambers. I spent those wonder years on the busy, public wards of Bellevue Hospital in the late 1950's where one soon learned that the drill-book we were following had been written in Boston. After the First World War, the Harvard Medical Unit at Boston City Hospital and its associated Thorndike lab had reformed the way medicine was taught and practiced in municipal teaching hospitals. Medical training programs countrywide soon followed Boston's dictum that laboratory and clinical experience were indispensable to each other, and both necessary for the training of young physicians. One could bring science from the bedside to the lab bench and back again—and also serve the medically needy. The Boston City/Thorndike model was a caring and curious doctor who could enjoy the luxury of doing good. The house-staff memoirs of the Harvard Medical Unit at Boston City Hospital at its zenith in the 1930's are a tribute to doctors and patients alike.[2] They are also the story of medicine before Medicare and Medicaid and penicillin.

In the thirties, Harvard was responsible for the Second and the Fourth of the five medical services at Boston City Hospital; the others were directed by Boston and Tufts Universities. Of these, perhaps the most prestigious was the Fourth, which was housed in the Peabody building, a structure connected to the Thorndike labs by "bridges" on which patients were housed in times of overcrowding. Boston City Hospital was an institution both charitable and commercial. On the one hand, the poor—"those incapacitated from taking care of themselves"[3] in the words of its 1860 prospectus—were treated at little or no cost to themselves; on the other, they paid their fee in the coin of medical instruction. Their illnesses became "cases," the raw material that medical students, house officers and attending physicians turned into the textbook of clinical medicine.[4] The best and the brightest of Harvard shuttled between those labs and the wards: the history of the Harvard Medical Unit at Boston City Hospital proudly announces that of the 71 young physicians who trained there between 1936 and 1940, 52 became professors of medicine, while 6 went on to the deanship of medical schools.[5]

Pups and Seniors

Interns at the Boston City Hospital in the 1930s signed up for an apprenticeship in internal medicine that would last eighteen months. The interns were expected to work their way up a steep ladder in six rotations of three months each. Beginning as a "pup," who was mainly expected to do the scut work, the novice rotated through a period of outpatient service, followed by another three months of tending to the infectious diseases ward. Then came the reward for those first nine months of dog's work: "the privilege of giving the orders instead of receiving them."[6] The intern became, in turn, a Senior Physician (the "senior" who looked after the pup), an Assistant House Physician, and finally, the House Physician. Nowadays, the "House" would be called the Chief Resident. There were no residents at Boston City in the modern sense of the term; after a year and one half of internship one simply went into practice or was called into the academy.

Nowadays critics of "paternalistic, hieratic" medicine complain that the patients were warehoused thirty or more to each open ward to serve the role of human guinea pigs. They tell us that the poor were piled bed upon bed like so much industrial inventory for the convenience of doctors, that charity cases provided the fodder of medical teaching.[4] But the memoirs of Finland, Castle, and their colleagues at Boston City Hospital paint an entirely different picture.[2] On the contrary, they show that the

Alexander Burgess, Bud Evans, and Lewis Thomas, Boston City Hospital (1937).

charity patients were the center of a busy hospital life, in which families, friends, clergy, doctors, nurses, medical students, interns, and custodians formed a community. In the days of the Great Depression, patients in the City Hospital found themselves in a precinct that was cleaner, warmer, and more caring by far than any slum in Boston. The records also show that in 1937, as in 1860, doctors and patients were parties to a barter agreement—care was given in exchange for teaching—a largely amicable contract that was unbreached for over a century before it was anulled by the HMOs :

> The hospital was part of an institutional world [in which] physicians were paid in prestige and clinical access; trustees in deference and the opportunity for spiritual accomplishment; nurses and patients were compensated with creature comforts: food, heat, and a place to sleep. Patients offered deference and their bodies as teaching material. Few dollars changed hands, but the system worked in its limited way for those who participated in it.[7]

Those directly involved had more visceral responses. Here are some diary entries of William L. Peltz, an intern on the Peabody Service of Boston City Hospital in the winter of 1938:

- To bed at 2:30 AM after working 18 1/2 hrs, one hour's sleep and now a new patient to be admitted at 3:30 AM. Age 55 diabetes with gastritis, says the supervisor's office. We'll see!
- Maria de Sista, aged eighteen with her TB on Peabody 2. She helped out with housework in the home of a woman who had TB. That was 4 years ago, and the woman has since died at Rutland [a TB sanitarium]. But Mary didn't know that at the time. A swell kid! Wanted to get married next April and now has advanced TB on the right. Came in spitting up blood. She began three months ago when she might have had a halfway decent chance, but some "friend" said not to worry, it was nothing, so she didn't bother going to the doctor . . .
- Anastasia O'Neil, who looks sixty-three but says she's seventy-four, who must weigh 250 pounds but says she's 150, who was sent in by her doctor with a note saying she has severe diabetes but who hasn't required insulin yet. "Sure an' they don't know what diabetes is in the old country; an' when yer sick the old folks know where to go out in the fields an' pick a big bunch o' herbs . . . Gee the hospital is a swell place; an' the City Hospital is as good as any place you can get . . .
- Diabetes! Anyway it is now 5:30. Now for two hours sleep—hopefully! (I am not up for alcoholics.)[8]

It's hard to remember now, but before the Second World War, medical services of the great teaching hospitals provided mainly custodial, rather than remedial, care: food, heat, and a place to get well. The doctors might relieve pain and suffering, but they had few real remedies in hand. They could give arsenic for syphilis, insulin for diabetes, raw liver for pernicious anemia, and antisera for pneumonia. But by and large, as one of them recalled "Whether you survived or not depended on the natural history of the disease. . . . And yet, everyone, all the professionals, were frantically busy, trying to cope, doing one thing after another, all day and all night.[9] Busiest of all were the interns.

Bill Peltz, who was to become a professor of psychiatry at Penn, described the intern's work as "typing blood, doing urinalyses and examining stools, giving transfusions, taking EKGs, typing pneumococci, pronouncing people dead, and signing death certificates."[8] The interns were also expected to work in the emergency room and the outpatient department from which they rushed back to the wards to start IVs, perform catheterizations, measure basal metabolism, do spinal taps, place tubes in various orifices "and more and more." Once past the pup stage, the interns admitted sick people at the rate of four or five per night, obtained their patient's social and medical histories, performed physical examinations, did all but the most difficult laboratory examinations, and, after mulling over all other possibilities, committed themselves to the single most likely diagnosis and plan of treatment. Then they waited. Over the next few hours, days, or weeks, they watched—but rarely influenced—the disease until the patient got better or worse; they were again required to keep detailed records of what happened. It was called "keeping"—or, later, "buffing"—the chart. Interns were also expected to comfort their patients, to make accurate prognoses, and when all had come to naught, to beg permission from the patient's nearest relative for an autopsy. In the course of these efforts they were expected to work every day and every other night; to ignore weekends—and to remain unmarried.

In return the interns received room and board and medical training that would last a lifetime. Franz J. Ingelfinger, who was Lewis Thomas's "senior" and later the best medical editor of his day, remembered looking up at the stars—or as much as could be seen between the Peabody building on one side and the House Officer's building on the other—and imploring the deities that he might do a good job: "it was an emotional and heady walk between those buildings."[10] Thomas, himself, had a fond recollection of the time:

> I am remembering the internship through a haze of time, cluttered by all sorts of memories of other jobs, but I haven't got it wrong nor am I romanticizing the experience. It was, simply, the best of times.[11]

Many of those who worked on the Peabody wards and at the Thorndike Memorial Laboratory agreed, "My experience on the Harvard Medical services was the most intense of my life," claimed a future professor of medicine in Seattle, Bud Evans.[12]

Nowadays, when broker's apprentices or bimbos in Washington are called "interns," we tend to forget that the title *interne* derives from the Parisian teaching hospitals of the middle of the nineteenth century who rewarded their best and brightest students with an *internat*. The losers in the intellectual lottery were awarded an *externat*.[13] Looking back at the memoirs of the Boston City days, I'm amused by the recent externalization of the internship. I remain convinced that the title of "intern" should be reserved for those who tend the sick *inside* a teaching hospital, those who live and work and sleep there, those who have enlisted to be on call, as one used to say.

Internship, then as now, had its pleasures as well as its trials. Five months into his internship Thomas celebrated in doggerel a gift that Mr. Maloof, a grateful fifty-three-year-old "Assyrian crock," gave to the doctors who had taken care of him. The gift was a curiously hammered brass pot which went to Peltz, who as House Physician had first call on gifts. Cary Peters (Assistant House) and Ingelfinger (the "senior"), who also appear in the poem, were further down the line of command. Thomas slipped into an office on Peabody 3 and typed an "Ode to a 90-Year-Old Assyrian Pot" which concluded:

And didst thou hope. Oh Burnished Pot
That such a fate should be thy lot?
That thou wouldst be so doubly blest:
To leave Maloof and be the guest
Of someone else?
Of William Peltz?
Oh happy Pot! Oh lucky toss!
Maloof came in while Peltz was boss!
For if he'd chose at home to tarry
Thou mights have gone to Peters, Carey!
And if he'd longer chose to linger
Thou mights have gone to Ingelfinger!
But no! Thou goest to no one else
But Dr. William Learned Peltz![14]

On Christmas morning of 1937, Thomas crossed over to the House Officer's building to post a note on his senior's door ; it had been written in the dead of night on Fourth Medical:

Of Christmas joys I am the Bringer;
I bring good news to Ingelfinger
Though many turned in bed and cried
Nobody died! nobody died![14]

A photograph has survived of these men sitting before the House Officers' building in the early fall sunshine of 1937.[15] Thomas is the senior of the group and is seated with Alexander Burgess and Bud Evans, his pup. They are three good-looking, neatly scrubbed young American doctors in crisp whites sporting white buck shoes, Thomas has a percussion hammer poking from the breast pocket of his starched tunic. The three men look at us across the years as if they knew, as Evans recalled, that the times had the attributes of the end of an era. It was certainly the end of a period of "do-it-yourself" activity in internal medicine.

> For instance, cross-matching blood at night and giving blood in kits sterilized by the house staff must have ceased shortly after this. Typing pneumococci and giving antipneumococcal serum during the night also adds glamour in retrospect that was removed when it became possible to prescribe sulfonamide and go to bed.[12]

And yet, that do-it-yourself experience became indelible. Richard Ebert, a future chairman of medicine at Minnesota, expressed it for all of them: "I developed a self-reliance and a knowledge of my clinical competence . . . the experience tested our abilities to withstand stress. The internship was a kind of indoor Outward Bound."[16] Those days come to life in Bill Peltz's casebook of 1937–1938 on the Peabody service. The sound is that of the Popular Front:

- Mannie Sample, the great brown colored girl in the Sun Room 1 whose toes I pinch and who has a laugh I like to hear.
- Catherine Healy who has diabetes and pyelitis and whom we have had in here twice now, damn near dead each time. Once with a tidal drainage going and the second time coming in insulin shock.
- Bessie Johnson, the little colored girl with pneumonia, who wrote me afterward as I hoped she would. I remember the night she was crying because she heard she was going to a convalescent home and how she called me over.
- Ventura Bocafuscia, the swell old Italian on the Peabody 1 bridge with his bigeminal pulse. He would teach me Italian each day. It was always a pleasure to make rounds past his bed because he was so cheerful. And his rhythm straightened out after quinidine.
- And Henry Coffin who . . . when he was operated on they found a huge ulcer too big to do anything for and he died a few days later.
- Going around on the Peabody 1 and 2 the last night and saying goodbye to all the patients and nurses and the ward help. Saying goodbye to Dr. [Soma] Weiss and the futile effort to express your sadness and regret.[17]

The Long White Coats

The house officers were under the spell of Soma Weiss and the other young Turks of the Harvard Unit, who

> moved back and forth between the Peabody building and their laboratories in the Thorndike . . . in long white coats. They came at ten in the morning to make the formal rounds, walked the bedsides for two or three hours with

> the interns and medical students, and they came back at odd hours throughout the afternoon and often until late in the evening to see patients with serious problems in whom they were especially interested.[6]

Chief of that glittering Harvard Medical Unit was George R. Minot (see "A Nobel Error"). It did not escape house-staff notice that one could be a teacher, a clinician, and a scientist all at once. Minot and his predecessor, Francis Weld Peabody, who had put the Thorndike together from scratch, combined each of those qualities. Scions of distinguished Boston families, they were also regarded by their house officers as avatars of the Brahmin physician. Interns on the Peabody service of the Boston City Hospital had each been given a copy of Peabody's *Doctor and Patient*; many of them knew key passages of the book by heart.[18]

Sporting a brisk introduction by Hans Zinsser, Peabody's *Doctor and Patient* provided generations of Harvard physicians with the rhetoric of clinical idealism—much in the way that William Osler's *Aequanimitas*[19] summed up the strictures of an earlier era. Peabody's book revealed the mind of a doctor who could link the art and science of medicine—and who could write about both. Peabody's dictum "the treatment of a disease may be entirely impersonal; the care of the patient must be completely personal" became a watchword for a generation of house-officers. Even more widely quoted was the last sentence of Peabody's essay "The Care of the Patient":

> The good physician knows his patients through and through, and his knowledge is bought dearly. Time, sympathy and understanding must be lavishly dispensed, but the reward is to be found in that personal bond which forms the greatest satisfaction of the practice of medicine. One of the essential qualities of the clinician is interest in humanity, for the secret of the care of the patient is in caring for the patient.[20]

At the time, Peabody was one of only two full-time professors of medicine at the Harvard Medical School.[21] Publication of his essay as the lead article in the *Journal of the American Medical Association* in 1927 and the simultaneous publication of his scientific work on pernicious anemia (see "A Nobel Error") constitute a milestone in the history of academic medicine in America. I remember reading the following passage early one morning in the chief resident's bunk at Bellevue Hospital in 1959 when I was wondering whether a career in full-time medicine would help or hinder my caring for patients:

> . . .The most common criticism made at present by older practitioners is that young graduates have been taught a great deal about the mechanism of

> disease, but very little about the practice of medicine or, to put it more bluntly, they are too "scientific" and do not know how to take care of patients. . . . When a patient enters a hospital, one of the first things that commonly happens to him is that he loses his personal identity. He is generally referred to, not as Henry Jones, but as "that case of mitral stenosis in the second bed on the left." There are plenty of reasons why this is so, and the point is, in itself, relatively unimportant; but the struggle is that it leads, more or less directly, to the patient being treated as a case of mitral stenosis, and not as a sick man. . . .[20]

When Peabody died, his shoes were filled by George Minot, and his path was followed by the young men he had brought to the Thorndike. *Primum inter pares* among these was the Hungarian-born Soma Weiss (1899–1942). Weiss's brief, meteoric career assumed some of the aura of Xavier Bichat, the founder of modern tissue pathology in the Napoleonic era.[21] Weiss grew up in a Hungarian mountain town, the son of a successful civil engineer. In response to the anti-Semitic turbulence of Budapest after World War I, Weiss arrived in New York in 1920 and is said to have applied at the same time, and with success, for American citizenship and admission to Cornell University Medical College, then located directly across the street from Bellevue Hospital. After internship at Bellevue, Weiss was attracted to the Thorndike by Peabody, where Peabody's Young Turks were busy "running a laboratory, teaching on the wards, and providing research training for young doctors. . . ."[6]

Weiss was an enthusiastic teacher, an unflagging clinical researcher, and an effective administrator. Among his various accomplishments were the first description of how patients fainted in the "carotid sinus syndrome" (demonstrated by projecting the shadow of the quartz filament of an old electrocardiograph on the front wall), the introduction of intravenous sodium amytal (first used to relieve the severe contractions of tetanus), and measurement of the circulation time at the bedside (by means of histamine or vital dyes). With George P. Robb, he first showed that paroxysmal nocturnal dyspnea (waking up short of breath in the middle of the night) was caused by overload of the left ventricle. These days his name remains in our literature as half of an eponym for the Mallory-Weiss syndrome: a rupture of the esophagus because of sustained vomiting, as after too much alcohol.[22] The house staff remembered him best as *the* man to call on, when on long winter nights there was a question to be answered, or a patient to be seen. He was "always about."

Weiss showed the same unflagging energy when he became chief of the Second and Fourth services. William Bean, a future professor at Iowa and a faithful scribe of medicine, remembers that Weiss's bedside demonstrations were models of clarity and often a virtuoso's tour de force. "His Hungarian accent, which lent charm to his almost dangerously attractive personality quite unconsciously intensified at certain times and almost vanished at others."[23] Alexander Langmuir, the future chief epidemiologist of the CDC, recalled that the thirty-four-year-old Weiss was constantly on the wards: "he was never dull. Then in a flash he was off, always running with his white coat tails flying."[24] He was deeply concerned about each patient and never let his interns forget the slightest detail of clinical management. He insisted that the interns not be diverted by the countless medical fascinations of the Boston medical scene, even by the excellent libraries, until they had worked up all their cases completely.

Said Weiss: "You will never have a chance to see human disease as it unfolds each hour day and night, under your eyes on the wards. You can hear distinguished lectures and read journals the rest of your life."[23] It was a restatement, for Boston, of Osler's famous dictum that he who observes patients without studying books is like one at sea without a chart. But he who studies books without observing patients is like one who has never been to sea at all.[24]

The rounds began early in the morning and often ended late at night when Weiss and his wife (Elizabeth Jones Weiss) entertained interns and fellows at home: "The Weiss's home was a happy holiday haven for homesick fellows," recalled Bean.[25] Paul Kunkel, a future Yale professor, recalled that one of the fondest memories he had of Weiss was the intimate Sunday evening buffets at home where he, his charming wife and their children presided. Invariably they had faculty from the liberal arts . . . who would keep conversation flowing at a lively pace.[26] Scores of Thorndike alumni remember best those Wednesday evening rounds, when the house staff of all the Boston City Hospital medical services, Boston University and Tufts well as Harvard, would present Soma Weiss with the toughest and most puzzling clinical cases. Kunkel remembers that: "Soma carried a small black notebook that he peeked at occasionally after examining the patient and while still listening to the presentation and I do not recall he was ever in error."[27] Often, after having come up with the correct diagnosis Weiss would end his remarks with "I've seen it again and again!"[28]

At age forty, Weiss was appointed the Hersey Professor of the Theory and Practice of Physic and Physician-in-Chief at the Brigham Hospital, Harvard's most prestigious academic chair. Sadly, Soma Weiss's time at Brigham was brief, because in December of 1941 he was stricken by a severe headache which he correctly diagnosed himself as a subarachnoid hemorrhage. One month later he was dead. The house officers of the day always regarded Weiss as the most engaged teacher they had ever met: those nights on the wards, the warmth of the Sunday buffets, the keen questions to the point; Weiss was one of those who made it the best of times for a young intern.

When I was chief resident at Bellevue in the fifties, not very much had changed: Minot and Peabody, Weiss and Peltz would have felt at home. The training was still a bit like boot camp and the indigent patients of our big city hospital were still warehoused on open wards. Antibiotics, diuretics, and cortisone had, of course, become available and the interns—the best and brightest of Bellevue—were able to do a good bit more for their patients, but young diabetics continued to be admitted with recurrent bouts of pyelitis, "little colored girls" came in with pneumonia, and grown men died of huge ulcers after futile operations. An intern's life at Bellevue in the fifties was like Boston City in the thirties: the constant busywork, those hours on open, overcrowded, and underheated winter wards, those nights spent looking at blood and urine, the intense drive to be of use. Those who worked on Lewis Thomas's service at Bellevue Hospital wouldn't have gotten it wrong either. For us it was also, simply, the best of times. But, Thomas would also have been the first to argue the point that nostalgia is no substitute for quick and certain knowledge. By the eighties Thomas knew that there was no going back:

> Medicine is no longer the laying on of hands, it is more like the reading of signals from machines . . . there is no changing this, no going back; nor, when you think about it, is there really any reason for wanting to go back. If I am in a bed in a modern hospital, worrying about the cost of that bed as well, I want to get out as fast as possible, whole if possible.[29]

Childish Curiosity

> The Eighteenth Century, of course, had its defects, but they were vastly overshadowed by its merits. It got rid of religion. It lifted music to first place among the arts. It introduced urbanity into manners, and made even war relatively gracious and decent. It took eating and drinking out of the stable and put them into the parlor. It found the sciences childish curiosities, and bent them to the service of man, and elevated them above metaphysics for all time.
>
> —H. L. Mencken, *The New Architecture*, 1931

THE TWENTY-FIRST CENTURY seems preoccupied with reversing each and every one of the accomplishments of the eighteenth. Indeed, in the United States these days, "childish curiosity"—the root of all good science—is more likely to be bent to the service of politics than to human needs. In no aspect of public life is the subversion of original science to bureaucratic need more evident than in the recurrent effort of the NIH to centralize the direction of biomedical research. First the doyens of D.C. produced a Byzantine "Roadmap" for experimental biology,[1] then they had a go at developing "a new discipline of clinical and translational sciences"[2] by creating an on-again, off-again empire of "translational research centers."

In keeping with the custom of their band, the central planners were marching to music written a generation ago. In 1974, Lewis Thomas complained that

> It is administratively fashionable in Washington to attribute the delay of applied science in medicine to a lack of planning. . . . Do we need a new system of research management, with all the targets in clear display, arranged to be aimed at?[3]

Thomas's "new system of research management, with all the targets in clear display" aptly describes the Roadmap and the Translational Research Centers of 2005.

We might remind the Centrists, as one could call them, that the discipline of clinical science is not new; it dates from 1908 when the American Society for Clinical Investigation was founded. We might also remind the Centrists that *all* research is translational. When W. T. Astbury of Leeds first defined molecular biology in 1961 as "predominantly three-dimensional and structural [but also] inquiring into genesis and function,"[4] he was translating the Bragg equation, $n\lambda=2d\mathit{sin}\theta$, into solid-state physics. In turn, the Bragg equation was itself a translation of Euclid's second theorem into Newton's *Opticks*. When Daniel J. McCarty used the Bragg equation to determine that crystalline, but not amorphous monosodium urate, caused gout he was translating physics into rheumatology,[5] and when $n\lambda=2d\mathit{sin}\theta$ was used to show that aging human aortas accumulate hydroxyapatite (bone), physics had been translated into gerontology.[6] New science indeed!

We ought to remind our leaders that if we want to nourish original science we need to support the childishly curious, not the politically astute. If it's the atomic bomb you want, start recruiting for the Manhattan Project. But if it's atomic theory you're after, look for the lone thinker who comes up with $e=mc^2$. You can plug through the base pairs of the human genome with a consortium or a company but you'll never come up with Erwin Chargaff's discovery that

> . . . the molar ratios of total purines to total pyrimidines, and also of adenine to thymine and of guanine to cytosine, were not far from 1.[7]

And while you can support an assembly line of hackers working their way through cancer SNP's, you'll never generate a hypothesis as prescient as Chargaff's:

> We must realize that minute changes in the nucleic acid, e.g., the disappearance of one guanine molecule out of a hundred, could produce far reaching changes . . . ; and it is not impossible that rearrangements of this type are among the causes of the occurrence of mutations."[7]

The work of Einstein, of the Braggs, of Chargaff—of most good scientists since the Enlightenment—has been the work of adults who permit themselves to follow their curiosity in the way children do. True amateurs of science, they discovered the new because of their love for the game.

Look at any dictionary—the Oxford English Dictionary will do—and you'll find two definitions of amateur; only one of them is complimentary. The first derives from the Latin *amare* (to love) and describes an amateur as

> one who loves or is fond of; one who has a taste for anything.

This kind of amateur has a choice, and could as easily love claret as the clarinet, or prefer mussels to fossils. In the heady days of the Enlightenment all of science was done by amateurs; indeed, "scientist" didn't supplant the term "natural philosopher" until William Whewell produced it upon the request of Coleridge in 1833.[8] Voltaire and his mistress Madame du Châtelet, those amateurs of natural philosophy, performed serious experiments in thermodynamics; John Dryden and Christopher Wren debated natural science with Robert Hooke and Robert Boyle before the Royal Society; Goethe learned about electricity from Jean Paul Marat. But then Lavoisier lost his head on the guillotine and soon enough natural philosophers were expected to become card-carrying professionals: the term "amateur" assumed its more common, negative connotation. The O.E.D.'s second definition of amateur is:

> One who cultivates anything as a pastime, as distinguished from one who prosecutes it professionally; hence, sometimes used disparagingly, as dabbler, or superficial student or worker.

In support of this definition, the O.E.D. continues by equating amateurism with "dilettantism." I'd bet that the Centrists are drafting their Roadmaps and plotting Translational Research Centers because they suspect that some scientist, somewhere out there, is pursuing science as a pastime, as an amateur in thrall to childish curiosity.

Let me introduce another example of why biomedical science does very well, thank you, without Roadmaps or Translational Research Centers. The newspapers are in high dudgeon about Merck's Vioxx debacle. It's important to note that this tangled story of billions gained and billions lost began with two curious amateurs of science. Prostaglandins were discovered seventy-five years ago by an obstetrician, Raphael Kurzrok, and a pharmacologist, Charles C. Lieb. Kurzrok wondered why a woman in Brooklyn had lower abdominal pain each time she had sexual intercourse and decided that his patient's pain was very much like the pain of uterine contraction during childbirth. Kurzrok and Lieb speculated that there might be a substance in human semen that caused the smooth muscle of the uterus to contract. Aided by Sarah Ratner, a fledgling biochemist, and after the usual trial and error, they were able to provide a partial validation of their theory. In 1931 they reported in the *Proceedings of the Society for Experimental Biology and Medicine* (not the trendiest of scientific journals) that when one centiliter of human semi-

Title page of Francesco Stelluti's *Melissographia*
(published by the Accademia dei Lincei, 1625)

nal fluid was added to a strip of human uterus suspended in a water bath, the uterine muscle sometimes contracted.[9] Later, in Sweden, the contracting substance of Kurzrok and Lieb was named prostaglandin by U.S. von Euler, since it was presumed to come from the prostate gland. Other Swedish scientists (Sune Bergström, Bengt Samuelsson) translated the story of prostaglandins into physiology, then into synthetic biochemistry, and finally into medicinal chemistry. Without a roadmap, the path of pure curiosity led straight to the Nobel Prize.

By 1971 the late Sir John Vane—another Nobelist with a grand streak sense of childish curiosity—announced "I think I know how aspirin works" and popped into the lab to show he'd got it right. Among other effects, aspirin and aspirin-like drugs do indeed inhibit the COX enzyme(s) of prostaglandin synthesis.[10] Vane's discovery was followed by Philip Needleman bumping into the COX 2 enzyme and distinguishing how its inhibitors differed from inhibitors of aspirin-like drugs with respect to gastric toxicity. The crystal structure of each COX enzyme was solved ($n\lambda=2d\sin\theta$, again) and ever more specific inhibitors were synthesized. What was being translated and in what direction? Chemistry took over from pharmacology what structural biology learned from gastroenterology which in turn borrowed from pathology, etc., etc.[11] It's all of modern science in one class of drugs!

Now that these adventures in translational research have been trumped by an unwanted, clinical effect of COX 2 inhibition, I've looked in vain on that Roadmap or Center directory for guidance. I'd bet that one lone, curious scientist, someone following up a hunch in another field (say cell-cell adhesion), will come up with the answer to why Vioxx is a plus/minus. And she'll get there before those chaps looking at Roadmap. But, we already have a counter to the best-laid plans of NIH mice and men, to the notion that protocols from above can direct our science. Here's Lewis Thomas again:

> What [research] needs is for the air to be made right. If you want a bee to make honey, you do not issue protocols on solar navigation or carbohydrate chemistry, you put him together with other bees . . . and you do what you can to arrange the general environment around the hive. If the air is right, the science will come in its own season, like pure honey.[12]

Jacques Loeb and Stem Cells

> Tonight I ask you to pass legislation to prohibit the most egregious abuses of medical research: human cloning in all its forms. . . . Human life is a gift from our Creator—and that gift should never be discarded, devalued or put up for sale.
>
> —George Bush, State of the Union Speech, January 31, 2006.[2]

Stem Cells and IVF Research

THE PRESIDENT'S HOSTILITY TO RESEARCH involving human zygotes has a long history. A generation ago, in 1971, Dr. Leon Kass anticipated Mr. Bush in a polemic against IVF entitled "Babies by means of in vitro fertilization: unethical experiments on the unborn?" Kass raised the banner of bioethics against assisted reproduction and defined a rigid notion of the origin of life.[3] In turn, Kass's question was posed by the devout at the dawn of the twentieth century, when Jacques Loeb at the Marine Biological Laboratory succeeded in activating a sea urchin egg in vitro: what price for life in a dish of salt water? The *New York Times* downplayed "Dr. Loeb's Incredible Discovery," calling it "very interesting but not especially important, [since it] will not revolutionize our concepts of the origin of life."[4]

But Loeb's sea urchin experiments became the basis of modern developmental biology both in the lab (think of cell cycles and cyclins) and in the clinic (think of Louise Brown, the world's first IVF baby, now twenty-seven years old). Meanwhile, Dr. Kass went on to chair President Bush's Council on Bioethics from 2001 to 2005. Undaunted by the worldwide success of IVF, and the thousands of families it has fulfilled, Dr. Kass and his Council raised the banner of bioethics again, this time against human embryonic stem cell research (hES). In concert with Congress and the administration, Kass and his Council have essentially scotched support for hES research by the federal government. Their stance has won strong support from the

National Review and the Discovery Institute, an organization devoted to beating up on Charles Darwin by teaching intelligent design in the schools.[5]

Swept by the tide of what Kevin Phillips has called the "American Theocracy" with its "rising commitment to faith as opposed to reason and a corollary downplaying of science,"[6] the Bush administration has restricted hES studies to a few unpromising cell lines established before August 9, 2001. Happily, individual states and private entities have assumed the task of new human stem cell research, but until recently the United States lacked a national set of standards, an ethical guide to the science, the hazards, and the promise of stem cell research.

ESCROCs

With courage and foresight, the National Academy responded to this need, and in April of 2005 issued its own "Guidelines for Human Embryonic Stem Cell Research" under the aegis of the NRC Board on Life Sciences and the IOM Board on Health Sciences Policy.[7] The NRC/IOM "Guidelines" deserve the support of everyone in science. This concise, yet definitive, 166-page document was drafted by an impeccable cast of scientists, lawyers, and clinicians: it constitutes a noble attempt to merge meliorist science with the temper of our time. In keeping with the model of DNA regulation by RAC's (Recombinant DNA Advisory Committees), the "Guidelines" sensibly called for letting a thousand oversight committees bloom, one for every institution that plans to conduct hES research. Other suggestions included a review of the procurement process, provisions for the banking of cell lines, assurance of informed consent, adherence to standards of clinical care, etc.—the document should serve for years to come as a guide to conduct and action.

One minor quibble: someone must have had a tin ear to propose the acronym "Embryonic Stem Cell Research Oversight (ESCRO)" committee for the local oversight bodies—ought we to pronounce the word "escrow," as in secure deposit? Worse yet, the complete acronym, ESCROC, should amuse anyone conversant with French: ESCROC means crook, as in a *Le Monde* headline of 1973: "Richard Nixon: *Je ne suis pas un escroc!*"[8]

But the awkward acronym is not the most puzzling aspect of ESCROCs; the restricted composition of their National Advisory committee is more of a problem. This spring, the NRC/IOM requested nominations for a super ESCROC to provide oversight of research with

Jacques Loeb[1]

human embryonic stem cells.[9] The "Guidelines" propose that "a national body should be established to assess periodically the adequacy of the policies and guidelines proposed in this document."[7] In keeping with the charge of its parent body, the call for nominations to the Advisory Committee carries the warning that "Please note that individuals conducting research using human embryonic stem cell will not be eligible (animal and adult stem cell researchers will be considered)." That exclusion seems to carry undertones of the president's 2006 State of the Union speech, or of Dr. Kass's 1971 warning against "unethical experiments on the unborn." One notes that Dr. Kass has remained as adamant as his President, testily reminding Nicholas Wade in 2005 of the party line on stem cells: Congress is still on record as saying, "federal funds may not be used in research in which human embryos are harmed or destroyed."[10]

Now, the NRC, as a private body, has every right to pick whomever it wants for its Advisory Council, and judging from the group that put together the "Guidelines" we can be assured of a concert-grade ensemble. Moreover, the staff has insisted that the exclusion of researchers who are actually working on hECs was by no means due to political pressure. The exclusion is in place, we are assured, "simply to avoid the conflict of interest that would arise if researchers working with human embryonic stem cells were to write the rules governing their own activities."[9] That argument isn't entirely persuasive, for at least two reasons that come immediately to mind. First: the RACs are populated by scientists who are themselves engaged in recombinant DNA research. Second: while there is no federal oversight body for IVF, its policy and practice are monitored by active members of the Society for Assisted Reproductive Technology. Its members see no conflict of interest between performing IVF themselves and the task of "setting and promoting the standards for the practice of assisted reproductive technology."[11] No, I'm afraid that the Academy's exclusion policy is a well-meaning display of caution and compromise in the face of implacable foes. A Freudian might call it a collective "introjection of the aggressor." I'm reminded that Jacques Loeb faced similar problems a century ago.

Loeb and In Vitro Fertilization

Each June I pass through a building named after Loeb at the Marine Biological Laboratory at Woods Hole, where reproductive biology has flourished since 1888. In 1900 Jacques Loeb was accused of transgressing the limits of science when he first produced viable sea urchin larvae by means of parthenogenesis.[12] Accusations flowed: fatherless babies violated natural law, divine intent had been breached, and Loeb charged with proposing the "atheistic hypothesis" that life had a physicochemical origin. He had, after all, created a form of new life in a dish. Loeb tried to calm the waters, perhaps in the same spirit of reasonable compromise shown by the Academies today: Loeb disclaimed in 1900 that "The experiments are not far enough developed to arrive at any definite conclusion regarding this subject, and I do not know how far we shall be able to go in the work of artificial production of life."[1] In the same vein, Loeb's colleague at the MBL, F. R. Lillie, assured a local newspaper that the creation of vertebrate embryos in vitro was a very long way off: "Why, it would be like finding the North Pole," he explained.[13] Admiral Peary found the North Pole nine years later.

Loeb also had his supporters. It was as clear to them in 1900, as to us today, that the main opposition to research on embryos in vitro came from "religious sources, where it is recognized that the general acceptance of this [materialist] hypothesis would overthrow the main doctrine, of an infinite all-powerful deity or creator, who has breathed the breath of life into every living thing."[15] Eventually, Loeb roused sufficient courage to answer a Jesuit critic that "one cannot overlook the fact that the steady progress of science dates from the day when Galileo freed science from the yoke of sterile scholastic methods."[16]

And sure enough on September 25th, 1912, the *Chicago Daily Tribune* announced that "PROF. LOEB HAS FATHERLESS FROG: Former Chicagoan Exhibits at Hygienic Congress Parentless Animal He Grew." The reporter chuckled in print that "by proper use of chemicals Prof Loeb was able to develop a mature frog, but alas, the professor, so learned in biology and chemistry did not know that a frog could not live in water and he let the poor thing drown. 'The next one will live,' says Prof Loeb, 'because I will bring him up on dry land.'"[14] It did, and the work went on.

From Loeb to Pincus and IVF

By 1935, Gregory Goodwin Pincus (1903–1967), best known as "The Father of the Pill," had worked out techniques for fertilizing mammalian eggs in vitro. An untenured assistant professor at Harvard, he also reported that he had activated rabbit eggs with sperm in vitro, reimplanted them in female rabbits, and obtained viable fetuses. Like Loeb, he was a champion of reductionist science: "Careful investigation of the ovum itself and its homeostatic environment is made possible by the various explantation and transplantation techniques."[17] We call it IVF.

The specter of human cloning has haunted the stem cell debate from the beginning. And by 1936, the specter appeared in the form of Gregory Pincus. Appreciating that "Not since Professor Jacques Loeb hatched fatherless sea-urchin larvae . . . has so striking a success been achieved," the *New York Times* warned its readers that Pincus's work would lead to embryo farms "making ninety-six human beings grow where only one grew before" as predicted in Aldous Huxley's *Brave New World*.[18] But in the 1930s, there were other specters. In rhetoric that smacked of Father Coughlin's nativist anti-Semitism, *Collier's* magazine published an attack on Pincus that featured an unflattering photo of this "Jersey native" as a cigarette-dangling mad scientist, holding a "fatherless" rabbit under his arm. The writer, J. D. Ratliff, traced Pincus's "goofy" experiments on in vitro fertilization to the work of "Jacques Loeb, the hugely famous Portugese Jew." Ratliff warned that with IVF "Man's value would shrink, the mythical land of the Amazons would then come to life. A world where woman would be self-sufficient; man's value precisely zero."[19] It was a popular view, in and out of the academy: Pincus was denied tenure at Harvard.

The fuss over Loeb and Pincus was enough to keep in vitro fertilization under wraps in the United States until 1978 (*pace* Landrum Shettles and John Rock). Meanwhile Britain got on with it and gave us Louise Brown: the model of less talk and more action. We've now had a replay of those early days of IVF in today's debate over hES. But, given that history, I'd bet that even our God-fearing lawmakers will prove powerless to stop scientists here and/or abroad from working on "the various explantation and transplantation techniques" (à la Pincus) required to bring human embryonic stem cell research from the dish into the clinic. And, thanks to the meliorist "Guidelines" of the Academies, the work will have a strong ethical base. One hopes that it will be overseen by an Advisory Committee staffed, in part, by those who've worked on human embryonic stem cells in lab and clinic.

Galton's Prayer

> What a tremendous stir-up your excellent article on prayer has made in England and America!
> —Charles Darwin to Francis Galton, November 8th, 1872[1]

AS THE EARTH WARMS, hurricanes, typhoons, and tsunamis happen. After each such calamity, be it in the land of Muslim, Jew, or Gentile, our leaders launch worldwide appeals for material aid and urge public prayer for the injured. New research shows that they needn't bother with prayer.

Well over a century ago, Sir Francis Galton, FRS, a cousin of Charles Darwin and the founder of modern biostatistics, called for a prospective, controlled study of whether those for whom prayers were offered would heal faster than those unaided by distant appeals to the deity.[2] Writing in the *Fortnightly Review* of August 1, 1872, Galton proposed the comparison of two groups of traumatically injured patients,

> The one consisting of markedly religious, piously-befriended individuals, the other of those who were remarkably cold-hearted and neglected [since] an honest comparison of their respective periods of treatment and the results would manifest a distinct proof of the efficacy of prayer.[2]

In 2006, 134 years later, his call was answered in a definitive, double-blind study published in the *American Heart Journal*, fetchingly named "The Study of Therapeutic Effects of Intercessory Prayer (STEP)."[3] Both its scope and cost were greater than the trial proposed in 1872.

STEP and Fetch It

Herbert Benson of Harvard and a brigade of faithful collaborators assigned three Christian prayer groups to pray for 1,800 patients undergoing coronary artery bypass graft (CABG) surgery in six medical centers throughout the United States. Funded mainly by the John

Templeton Foundation, which supports research at the religion/science interface, the $2.4 million study was touted as "the most intense investigation ever undertaken of whether prayer can help to heal illness."[4] It found that patients undergoing CABG surgery did no better when prayed for by strangers at a distance to them (intercessory prayer) than those who received no prayers. But 59 percent of those patients who were told they were definitely being prayed for developed complications, compared with 52 percent of those who had been told it was just a possibility, a statistically significant, if theologically disappointing, result. Benson *et al.* came to the objective conclusion that "Intercessory prayer itself had no effect on complication-free recovery from CABG, but certainty of receiving intercessory prayer was associated with a higher incidence of complications."

American advocates of intercessory prayer immediately raised objections to the work. "'GOD FACTOR' DEFENDED; PRAYER STUDY FLAWED" headlined the Worcester *Telegram and Gazette*. Brother Dennis Anthony Wyrzykowski told the newspaper that "The study was not a ploy to make God look bad," he said. "Dr. Benson is interested in the God factor. He's not out to disprove anything. We were disappointed. I was sure it would show that prayer works. . . ."[5] "AREA RESIDENTS CHALLENGING PRAYER STUDY" headlined the *Richmond Times-Dispatch*. Rev. Robert Friend, of All Saints Episcopal Church argued, "That's not the way prayer works. When we pray we are aligning ourselves the best we can with God's will for us. I think God's will for us is that we be whole and healthy." All Saints will continue its weekly prayer service for healing.[6]

Newspapers in less devout corners of the earth drew different conclusions: "HEALING POWER OF PRAYER DEBUNKED" cried the *Gazette* of Montreal, while *Le Monde* warned from Paris that "*LA PRIÈRE SERAIT DANGEREUSE POUR LA SANTÉ*" ["Prayer is dangerous to your health"]. Britain's *The Guardian* called attention to the most unexpected result of the STEP study: "IF YOU WANT TO GET BETTER—DON'T SAY A LITTLE PRAYER." *The Guardian's* Oliver Burkeman wrote, "If a religious person offers to pray for you next time you fall ill, you may wish politely to ask them not to bother. The largest scientific study into the health effects of prayer seems to suggest it may make matters worse."[7]

Portrait of Sir Francis Galton, FRS, 1822–1911
at age fifty-six

Prayers and Royals

The Guardian didn't need Benson's study to draw its conclusions. Indeed, the STEP trial had been scooped by an older, irrefutable, and far less expensive analysis: Galton's own essay of 1872. Galton suggested that in place of any prospective study of intercessory prayer in traumatic injury, a more rigorous test of the effect of prayer would require life or death as an endpoint. He proposed an inquiry

> into the longevity of persons whose lives are prayed for; also that of the praying classes generally; and in both those cases we can easily obtain statistical facts.[2]

He noted that, from time immemorial, public prayer each Sunday has been offered from the *Book of Common Prayer* for the long life of the Royal Family of England as in "Grant him/her in health long to live." He asked "Now, as a simple matter of fact, has this prayer any efficacy?" Fetching data from the records of over 6,500 biographies assembled by a colleague, Galton found that intercessory prayer offered on behalf of those who knew they were prayed for was bad for one's health.

It turned out that members of royal houses had a mean life expectancy of 64.0 years, significantly less than that of other aristocrats (67.3 years) or of other gentry (70.2 years). Since the royals knew they were being prayed for—we might say that each Sunday they were on the same page as their subjects—they were as much at risk for a worse outcome from prayer as were Dr. Benson's CABG patients. While Benson *et al.* downplayed the possible side effects of prayer as a possible chance finding, Galton had reached a similar conclusion from a neater endpoint.

Worse yet! The life expectancy of eminent clergymen (66.4 years) was no higher than that of their peers among lawyers (66.5 years) or medical men (67.0 years) and Galton concluded that, in keeping with the results of the STEP trial,

> prayers of the clergy for protection against the perils and dangers of the night, for protection during the day, and for recovery from sickness, appear to be futile in result.[2]

In an editorial that accompanied the STEP trial, William Krucoff and colleagues at Duke sounded concern over the higher incidence of complications in the group that knew it was being prayed for. They asked: "If the results had shown benefit rather than harm, would we have read the investigators' conclusion that this effect 'may have been a chance finding,' with absolutely no other comments, insight, or even speculation?"[8] Supported by a Foundation committed to the "God factor" at work in health and disease, Benson *et al.* must have been taken aback by the harmful consequences of intercessory prayer. Perhaps that's why it took almost five years to analyze data obtained on patients enrolled in the trial from January 1998 to November 2000 (!)[9] Undaunted, Dr. Charles Bethea, one of STEP's coauthors, insisted that "One conclusion from this is that the role of awareness of prayer should be studied further."[4] While the final STEP publication was impartial in its presentation and low-keyed in interpretation, it aroused the anger of believers and skeptics alike. The credulous contended that STEP was flawed, that it represented bad medical care and trivialized religion.[7] Skeptics argued against the study, convinced that there is no place in the realm of science for supernatural intervention.[10]

STEPS to Endarkenment

Now, it is certainly within the prerogative of objective clinicians to engage in statistical analyses of long-range prayer for others, especially when a Foundation devoted to such notions picks up the tab. On the other hand, not only skeptics will wonder why the National Institutes of Health would encourage or support inquiries into the supernatural. Newspaper accounts of the STEP trial carried the remarkable news that our government has spent more than $2.3 million of public money on prayer research since 2000.[4] Some of these studies overlap the published results of Galton and Benson *et al.*, including, for example, "Distant Healing Efforts for AIDS by Nurses and 'Healers'" (Elisabeth F. Targ, PI California Pacific Medical Center, 1-R01-AT-485-1), a three year grant totaling approx $663,000 and "Efficacy of Distant Healing in Glioblastoma Treatment" (*idem* PI, 1-R01-AT-644-1), a four-year grant totaling approximately $823,000. Results of these studies have not yet been made public, but as Martin Gardner has reported, the principal investigator on these grants has long been a fan of distant healing. In April of 2000,[11] she reported that

> Of more than 135 studies of distant healing on biological organisms . . . about two-thirds reported significant results. One fascinating study . . . concerned remote healing of tumors on mice. *The study showed that the healers who were farthest from the mice had the greatest influence in shrinking the tumors* [my italics].[12]

To be fair, when conducting controlled trials, Dr. Targ has been as professional and objective as Dr. Benson and his colleagues in their STEP study. Reporting on an $800,000 trial funded by the Department of Defense (Grant No. 17-96-1-6260) of Complementary and Alternative Methodologies (CAM) in breast cancer, "The efficacy of a mind-body-spirit group for women with breast cancer" she concluded that "The study found equivalence on most psychosocial outcomes between the two interventions" (CAM vs control).[12] And that's not even at a distance.

This year, the chances of being funded on any given grant application to the NIH may fall well below 10.0%.[13] It is in this context that one questions whether the NIH, and especially the National Center for Complementary and Alternative Medicine (NCCAM) has any business encouraging further grant applications and/or research into prayer. Those of us who engage in experimental biology are generally uninterested in enlarging the norms of our realm into the spiritual, artistic or ethical life of our time. But believers in "noetic," spiritual, or supernatural explanations for the vast territory of the unknown in science seem to have no such qualms. They've persuaded a credulous citizenry that there is spiritual gold to be mined by applying the methods of science to the study of religious practice. By confusing credulity with piety, they've also cleared the way to belief in "intelligent design." While such notions discredit both rigorous science and true beliefs, they are part and parcel of the new Endarkenment.

Twenty years ago, when our nation's scientific repute in science was at apogee, the United States accounted for about 40 percent of the total number of reputable scientific papers published in the world, the European Union for 33 percent and the Asia-Pacific region for 14 percent.[14] No longer: the seats of American power have been usurped by fans of unreason who feel free to preach "creation science," "alternative" medicine, "faith-based" social service, and bible-thumping homophobia. In consequence, the standing of American science has been eroded. By 2004, the EU had moved into the lead with 38 percent of total scientific papers published worldwide, the United States had slipped to 33 percent, while the Asia Pacific region moved up rapidly to become the source of 25 percent of all papers.[15] It's hard to see how federal action to prevent flag-burning, to ban gay marriage, or to hinder research on stem cells can reverse this trend.

NCCAM is undeterred. Catherine Stoney, PhD, of its Division of Extramural Research and Training insists that: "There is already some preliminary evidence for a connection between prayer and related practices and health outcomes. For example, we've seen some evidence that religious affiliation and religious practices are associated with health and mortality—in other words, with better health and longer life."[15] She is unlikely to have consulted Galton's statistics in *The Fortnightly Review.*

Galton, who shared Erasmus Darwin as a grandfather with Charles Darwin (all three were Fellows of the Royal Society) was far more modest in his peroration, as he offered the skeptical equivalent of prayer:

> Neither does anything I have said profess to throw light on the question of how far it is possible for man to commune in his heart with God . . . and it is equally certain that similar benefits are not excluded from those who on conscientious grounds are skeptical as to the reality of a power of communion. . . . They know that they are descended from an endless past, that they have a brotherhood with all that is, and have each his own share of responsibility in the parentage of an endless future. The effort to familiarize the imagination with this great idea has much in common with the effort of communing with a God, and its reaction on the mind of the thinker is in many important respects the same. It may not equally rejoice the heart, but it is quite as powerful in ennobling the resolves, and it is found to give serenity during life and in the shadow of approaching death.[2]

References

Introduction: The Endarkenment

1. D. Kennedy, "Twilight for the Enlightenment?" *Science* 308(5719):165 (April 8, 2005).

2. http://www.gallup.com/poll/content/login.aspx?ci=14107

3. J. Monod, *Chance and Necessity*, translated by Austryn Wainhouse (New York: Knopf, 1971), p.170.

4. Christoph Schonborn, "Finding Design in Nature" [Op-Ed], *New York Times,* July 7, 2005, p. A23.

5. NAS Press 1999, "Science and Creationism: A View from the National Academy of Sciences," http://books.nap.edu/html/creationism/index.html

6. G. B. Dalrymple, *The Age of the Earth* (California: Stanford University Press, 1991).

7. L. L. Cavalli-Sforza and F. Cavalli-Sforza, *The Great Human Diasporas: The History of Diversity and Evolution* (New York: Perseus Books, 1995), p. 40.

8. *Ibid.*, p. 51.

9. T. A. Schlenke and D. J. Begun, "Natural selection drives Drosophila immune system evolution," *Genetics* 164(4):1471–80 (August 2003).

10. R. Hirschhorn, D. R. X. Yang, J. M. Puck, M. L. Huie, C. K. Jiang, L. E. Kurlandsky "Spontaneous in vivo reversion to normal of an inherited mutation in a patient with adenosine deaminase deficiency," *Nature Genetics* 13(3):290–295 (1996).

11. M. Gross, H. Hanenberg, S. Lobitz, *et al.*, "Reverse mosaicism in Fanconi anemia: natural gene therapy via molecular self-correction," *Cytogenet Genome Res.* 98(2–3):126–135 (2002).

12. J. V. Ruderman, V. Sudakin, and A. Hershko. "Preparation of clam oocyte extracts for cell cycle studies," *Methods Enzymol.* 283:614–22 (1997).

Intelligent Design: Galileo and the Lynxes

1. D. Sobel, *Galileo's Daughter* (New York: Penguin, 2000), p. 274–5.

2. R. Hooke, Preface to *Micrographia* (1665; Reprint, New York: Dover, 1962), p. xi.

3. D. Stout, "Frist Urges Two Teachings on Life Origin," *New York Times,* August 20, 2005, p. A10.

4. http://caselaw.lp.findlaw.com/cgi–bin/getcase.pl?court=us&vol=393&invol=97

5. Sobel, p. 275.

6. D. Freedberg, *The Eye of the Lynx: Galileo, His Friends, and the Beginnings of Modern Natural History* (Chicago: University of Chicago Press, 2002), p. 123.

7. S. Drake, *Discoveries and Opinions of Galileo* (New York: Anchor Books, 1957), p. 85.

8. Horace, *Odes and Epodes*, translated by C. E. Bennett, Loeb Classical Library, (Cambridge: Harvard University Press, 1999), p. 176. [G.W. transl.].

9. http://www.constitution.org/bacon/instauration.htm

10. http://www.thelatinlibrary.com/horace/epist1.shtml [G.W. transl.].

11. Hooke, p. xi.

12. G. F. Bignami, "The microscope's coat of arms . . . or, the sting of the bee and the moons of Jupiter," *Nature* 405:999 (2000).

13. http://brunelleschi.imss.fi.it/apiarium

14. Hooke, p. 106.

15. http://www2.sas.ac.uk/warburg/pozzo/fossil.html

GALILEO'S GOUT:

1. P. N. Spotts, "For Galileo, one last voyage of discovery; Scientists hope to retrieve more data from a craft that has already left an indelible mark on planetary research," *Christian Science Monitor*, November 8, 2002, p. 3.

2. "*Eppur si muove*—or maybe not; Europe's Galileo satellite positioning system," *The Economist*, London, June 1, 2002, p. 47.

3. G. Johnson, "Here They Are, Science's 10 Most Beautiful Experiments," *New York Times*, September 24, 2002, p. 21.

4. D. Freedberg, *The Eye of the Lynx: Galileo, His Friends, and the Beginnings of Modern Natural History* (Chicago: University of Chicago Press, 2002).

5. G. Galileo, "The Assayer," in *Discoveries and Opinions of Galileo* S. Drake, ed. (New York: Anchor Books, 1957) p. 275.

6. A. Clements, "Philip Glass's Embarrassing Opera: *Galileo Galilei*: Barbican, London" *The Guardian*, Manchester (UK), November 2, 2002, p. 16.

7. http://www.pbs.org/wgbh/nova/Galileo

8. T. Walsh, "Exploring a Mind–and the Heaven," *Boston Globe*, October 27, 2002.

9. Freedberg, p.129.

10. D. Sobel, *Galileo's Daughter* (New York: Walker, 1999).

11. http://es.rice.edu/ES/humsoc/Galileo/Villa/galileo_pictures.html

12. R. Porter and G. S. Rousseau, *Gout: The Patrician Malady* (New Haven: Yale University Press, 1998).

13. C. H. Espinel "Michelangelo's gout in a fresco by Raphael," *The Lancet* 354(9196):2149–51 (December 18–25, 1999).

14. G. de Baillou, *Opuscula medica, de arthritide, de calculo et de urinarum hypostasi*, Editore M. Jacobo Thevart.), Paris: J. Quesnel. 1643, p. 199 (*quoted* in Porter and Rousseau *above*, p. 42).

15. Porter and Rousseau, p. 72.

16. G. Weissmann, "A Fashion in Metals" in *The Woods Hole Cantata* (New York: Dodd, Mead, 1985) pp. 91–101.

17. A. Antonello, M. Rippa Bonati, A. D'Angelo, G. Gambaro, L. Calo, L. Bonfante, "Gout and kidney during XVII and XIX centuries," *Reumatismo* 54(2):165–71 (April–June 2002).

18. S. Kotagal, "Episodic headache as a manifestation of lead encephalopathy," *Headache* 22(4):189 (July1982).

19. S. Ling, C. Chow, A. Chan, K. Tse, K. Mok, S. Ng, "Lead poisoning in new immigrant children from the mainland of China," *Chin. Med. J.* (England) 115(1):17–20 (January 2002).

20. K. Jongnarangsin, S. Mukherjee, M. A. Bauer, "An unusual cause of recurrent abdominal pain," *Am. J. Gastroenterol.* 94(12):3620–2 (December 1999).

21. M. E. J. Curzon and B. G. Bibby, "Effect of heavy metals on dental caries and tooth eruption," *J. Dent. Child.* 37:463–465 (1970).

22. S. Kolb, S. Domschke, H. J. Konig and W. Domschke, "Lead poisoning—also today

still current," *Med. Wchshrft.* 24;99(36):1464–9 (September 1981).

23. C. Roggi, C. Minoia, A. Ronchi, A. Gatti, A. Mastretti, L. Maccarini, F. Meloni, C. Meloni, "Epidemiological study on alcohol consumption trends and on the effects of alcohol consumption on the human body," *Note 2:* "Levels of lead in red wine from a northern Italian region,"*Ann. Ig.* 5(2):97–105 (March–April 1993).

24. http://es.rice.edu/ES/humsoc/Galileo/MariaCeleste/

Swift-boating Darwin: Alternative and Complementary Science

1. L. Goodstein and K. Chang, "Issuing Rebuke, Judge Rejects Teaching of Intelligent Design," *New York Times,* December 21, 2005, p. A1.

2. Michael Powell, "Judge Rules Against 'Intelligent Design'" *Washington Post,* December 25, 2005, p. A3.

3. "Revolution in Evolution Is Underway, Says Thomas More Law Center," January 18, 2005. http://releases.usnewswire.com/GetRelease.asp?id=41768A

4. R. Crowther, "Intelligent Design" (Letter) *New York Times,* December 23, 2005.

5. www.discovery.org/csc/

6. http://highwire.stanford.edu

7. J. Polkinghorne, "Intelligent Design, Creationism and Its Critics. Philosophical, Theological and Scientific Perspectives," (Review) *J. Theol. Studies* 54: 460–461 (April 2003).

8. Rosenthal Center celebrates tenth anniversary http://www.cumc.columbia.edu/news/in-vivo/Vol2_Iss21_dec22_03/around_and_about.html

9. National Center for Complementary and Alternative Medicine http://nccam.nih.gov

10. Nicholas D. Kristof, "The Hubris of the Humanities" (Op-Ed), *New York Times,* December 6, 2005.

11. M. A. Zehr, "School of Faith," *Education Week*: 25 A27, December 7, 2005. file://localhost/11. http/::www.faseb.org:opa:pdf:EvolutionStatement.pdf

12. U.S. Bureau of Labor Statistics http://www.bls.gov/oco/oco20052.htm; http://www.adherents.com/Na/Na_41.html

13. T. Dobzhansky, "Nothing in biology makes sense except in the light of evolution," *The American Biology Teacher* 35:125–129 (1973).

14. D. Diderot, *"Eléments de physiology," [cited in]* J-P. Changeux, Catalogue of exhbitioni at Nancy, September–16 December 2005 *"La lumière au siècle des Lumières et aujourd'hui"* (Paris: Odile Jacob, 2005) p.16. http://www.google.com/search?q=cache:sp8025MO1z4J:www.college-de-france.fr/media/com_cel/UPL55874_dossier_expo_final.pdf+C'est+qu'il+est+bien+difficile+de+faire+de+la+bonne+métaphysique+et+de+la+bonne+morale+sans+être+anatomiste,+naturaliste,+physiologist

15. L. Huxley *The Life and Letters of Thomas H. Huxley* (New York: D. Appleton, 1901) p. 199.

16. R. Darnton, *Mesmerism and the End of the Enlightenment in France* (Cambridge: Harvard University Press, 1968).

17. R. Darnton, "Franz Anton Mesmer."Dictionary of Scientific Biography 9:325–*8,* (1974).

18. *"le fluide sans l'imagination est impuissant, alors que l'imagination sans le fluide peut produire les effets que l'on attribue au fluide."* http://www.science-et-magie.com/sm50/sm0003mesm.htm

19. Reel Classics: Danny Kaye http://www.reelclassics.com/Actors/Kaye/kaye.htm

Homeostasis and the East Wind

1. W. B. Cannon, *The Way of an Investigator: A Scientist's Experience in Medical Research* (New York: W.W. Norton, 1945), p. 19.

2. S. Freud, *An Autobiographical Study* (1925). Reprinted in *The Standard Edition of the Works of Sigmund Freud*, vol. 20 J. Strachey *et al.*, eds. (London: Hogarth Press, 1953) p. 52.

3. W. James, *The Letters of William James* (Boston: Atlantic Monthly Press,1930), p. 327.

4. S. Freud, *op. cit.*, p. 56.

5. G. W. Allen, *William James (*New York: Viking/Compass, 1969), p. 490.

6. James, *Letters* p. 328.

7. *Ibid.*, p.103.

8. W. James, "Reflex Action and Theism" in *Essays on Faith and Morals,* (New York: R.B. Perry, ed., New American Library, 1974), p. 131.

9. Allen, pp. 129–151.

10. S. Freud, *Beobachtungen über Gestaltung und feineren Bau der als Hoden beschriebenen Lappenorgane des Aals',* S. B. Akad. Wiss. Wien *(Math.-Naturwiss. KJ.*). 3: 227–230 (1877).

11. S. Freud, "*Beitrag zur Kenntnis der Cocawirkung," Wien. Med. Wschr.* 31:129 *(*1885).

12. F. Sulloway, *Freud: Biologist of the Mind* (New York: Basic Books, 1979), p. 14.

13. L. Menand, *The Metaphysical Club* (New York: Farrar, Straus & Giroux, 2001).

14. Allen, p. 372.

15. James, Letters to Alice James, December 24, 1885.

16. Allen, p. 271.

17. http://www.survivalafterdeath.org/articles/james/impressions.htm

18. Holmes-Pollock Letters, *The Correspondence of Mr. Justice Holmes and Sir Frederick Pollock 1874-1932,* edited by Mark DeWolfe Howe, vol. 2 (Cambridge: Harvard University Press, 1941), p. 103.

19. S. Freud, *Psychopathology of Everyday Life,* reprinted in *The Standard Edition of the Works of Sigmund Freud,* vol. 6; J. Strachey *et al.,* eds., (London: Hogarth Press, 1953–1964) p. 47

20. H. V. Helmholtz, "Aim and progress of physical science," in *Popular lectures on scientific subjects*, English trans. E. Atkinson (New York: Continuum, 1999 reprint of 1873 ed.), p. 384.

21. S. A. Benison, C. Barger, C. and E. L. Wolfe, *Walter B. Cannon: The Life and Times of a Young Scientist* (Cambridge: Harvard University Press, 1987), p. 258.

22. Benison, p. 259.

23. W. B. Cannon and D. de la Paz, *Am. J. Physiol.* 27:64–7 (1911).

24. W. B. Cannon and D. de la Paz, "The stimulation of adrenal secretion by emotional excitement," *JAMA* 56: 742–5 (1911).

25. D. Diderot, *D'Alembert's Dream*, 1769, repr. (New York: Penguin Classics, 1976), p. 155.

26. C. Bernard, *Cahier Rouge* (1855) Hebbel H Hoff, Lucienne Guillemin and Roger Guillemin trans. (Cambridge, Massachusetts: Shenkman Publishing,1967), p. 58.

27. W. B. Cannon in A. Pettit, ed. *Homage à Charles Richet* (Paris: *Presse Institut,* 1926), p. 9.

28. C. Richet, *Dictionnaire de Physiologie* (Paris: Felix Alcan Pub., 1900), p. 72.

29. W. B. Cannon, *Bodily changes in pain, hunger, fear, and rage; an account of recent researches into the function of emotional excitement* (New York, London: D. Appleton and Company, 1915).

30. W. B. Cannon, *The Wisdom of the Body* (New York: W. W. Norton, 1932).

31. W. B. Cannon, *The Way of an Investigator: A Scientist's Experience in Medical Research* (New York: W. W. Norton, 1945).

32. *Ibid.*, p. 68.

33. Benison, p. 405.

34. J. M. Brinnin, *The Third Rose: Gertrude Stein and Her World* (Boston: Little, Brown, 1959), p. 30.

35. W.B. Cannon, "Closing Remarks" in *Exercises celebrating twenty-five years as George*

Higginson Professor of Physiology (Cambridge: Harvard University Press, October 15, 1931), p. 64.

36. Horace, Book III, Ode 2, in *Odes* (G.W. transl.) (Cambridge: Loeb Classical Library/ Harvard University Press, 1969).

37. W. James, "The Moral Equivalent of War" (1910) in *Essays on Faith and Morals*, ed. R. B. Perry (New York: Meridian/New American Library, 1962), pp. 311–328.

38. W.B. Cannon, "The Physiologic Equivalent of War," *JAMA*, 63:1415–16 (1914).

39. W. Owen, "Dulce et Decorum est" in *Collected Poems* (London: Chatto & Windus, 1968), p. 132.

40. W. B. Cannon, *Traumatic Shock* (New York: Appleton, 1923).

41. Cannon, *The Way of an Investigator*, p.165.

42. *Ibid.*, p. 162.

43. D. Fleming, "Walter B. Cannon and Homeostasis," *Social Research* 51:609–640 (1984).

44. W. B. Cannon, *The Way of an Investigator*, p. 174.

45. E.L. Wolfe, A.C. Barger, S. Benison, *Walter B. Cannon, Science and Society* (Cambridge: Harvard University Press, 2000), p. 359 ff.

46. O. Loewi, "An Autobiographical Sketch," *Perspectives in Biology and Medicine* 4:3–25 (1960).

47. http://www.nobel.se/medicine/laureates/1936/loewi-bio.html

48. L. Lorand [personal communication].

49. W.B. Cannon, "Closing Remarks," p. 64.

Red Wine, Ortolans, and Chondroitin Sulfate

1. A. J. Liebling, *Between Meals* (New York: Simon & Schuster, 1962), p. 87.

2. A. Waugh, *In Praise of Wine and Certain Noble Spirits* (New York: William Sloane, 1959), p. 221.

3. http://globalis.gvu.unu.edu/indicator_detail.cfm?IndicatorID=18&Country=FR-

4. G.-M. Benamou, *Le Dernier Mitterand* (Paris: Editions Plon, 2005), p. 256.

5. http://www.bbr.com/US/db/newsitem/653?ID=null&first_news_F=1

6. S. Renaud and M. de Lorgeril, "Wine, alcohol, platelets, and the French paradox for coronary heart disease," *The Lancet* 339:1523–6 (1992).

7. S. Renaud, R. Gueguen, "The French paradox and wine drinking," *Novartis Found. Symp.* 216:208–217 (1998).

8. D. O. Clegg, D. J. Reda, C. L. Harris, *et al.*, "Glucosamine, chondroitin sulfate, and the two in combination for painful knee osteoarthritis," *N. Engl. J. Med.* 354(8):795–808 (February 23, 2006).

9. P. Rozin, C. Fischler, S. Imada, *et al.*, "Attitudes to food and the role of food in life in the U.S.A., Japan, Flemish Belgium and France: possible implications for the diet-health debate," *Appetite* 33:163–80 (1999).

10. J-A Brillat-Savarin, *Physiologie du goût, Médiation IV, de l'appétit* (Paris: Flammarion,1993 edition of 1826 orig.), p. 236.

11. S. Pervaizii, "Resveratrol: from grapevines to mammalian biology," *The FASEB Journal* 17:1975–1985 (2003).

12. M. Viswanathan, S. K. Kim, A. Berdichevsky, L. A. Guarente, "Role for SIR-2.1 regulation of ER stress response genes in determining C. elegans life span," *Dev. Cell.* 9:605–15 (2005).

13. A. S. Levenson, B. D. Gehm, S. T. Pearce, J. Horiguchi, L. A. Simons, J. E. Ward 3rd, J. L. Jameson, V. C. Jordan, "Resveratrol acts as an estrogen receptor (ER) agonist in breast cancer cells stably transfected with ER alpha," *Int. J. Cancer* 104:587–96 (2003).

14. G. Weissmann, P. Davies, K. Krakauer, R. Hirschhorn, "Studies on lysosomes. 13. Effects of stilbamidine and hydroxystilbamidine on in vitro and in vivo systems," *Biochem. Pharmacol.* 19:1251–61 (1970).

15. P. Marambaud, H. Zhao, P. Davies, "Resveratrol promotes clearance of Alzheimer's disease amyloid-beta peptides," *J. Biol. Chem.* 280:37377–82 (2005).

16. J. K. Kundu, Y. K. Shin, Y. J. Surh, "Resveratrol inhibits phorbol ester-induced expression of COX-2 and activation of NF-kappaB in mouse skin by blocking IkappaB kinase activity," *Carcinogenesis* (2006) [E-pub ahead of print].

17. J. J. Moreno, "Resveratrol modulates arachidonic acid release, prostaglandin synthesis, and 3T6 fibroblast growth," *J. Pharmacol. Exp. Ther.* 294:333–8 (2000).

18. B. N. Cronstein, M. C. Montesinos, G. Weissmann, "Salicylates and sulfasalazine, but not glucocorticoids, inhibit leukocyte accumulation by an adenosine-dependent mechanism that is independent of inhibition of prostaglandin synthesis and p. 105 of NFkappaB," *Proc. Nat'l Acad. Sci.U S A.* 96:6377–81 (1999).

CORTISONE AND THE BURNING CROSS

1. M. M. Hargraves, R. Morton, "Presentation of 2 bone marrow elements—the tart cell and the LE cell," *Proc. Staff M Mayo Clin.* 23(2):25–28 (1948).

2. H. M.Rose, C.Ragan, E.Pearce, *et al.*, "Differential agglutination of normal and sensitized sheep erythrocytes by sera of patients with Rheumatoid Arthritis," *Proc. Soc. Exp. Biol. Med.* 68(1):1–6 (1948).

3. W.A. Laurence, "Aid in Rheumatoid Arthritis Is Promised by New Hormone," *New York Times,* April 21, 1949, p. 1.

4. P. S. Hench, E. C. Kendall, C. H. Slocumb, and H. F. Polley, "The effect of a hormone of the adrenal cortex (17 hydroxy-11 dehydrocorticosterone: compound E) and of pituitary adrenocorticotropic hormone on rheumatoid arthritis," [Preliminary report] *Proc. Staff Meetings Mayo Clinic* 24:181–297 (1949).

5. http://www.nobel.se/medicine/laureates/1950

6. Anon. "Arson Fails at Home of Negro Scientist," *New York Times,* November 23, 1950, p. 29.

7. P. L. Julian, E. W. Meyer, W. J. Karpel, *et al.*, "Sterols 11. 17-Alpha-Hydroxy-11-Desoxycorticosterone (Reichstein Substance-S) *J. Am. Chem. Soc.* 72 (11):5145–5147 (1950).

8. http://www.nobel.se/medicine/laureates/1950/hench-lecture.pdf.

9. L. H. Sarrett, "The partial synthesis of dehydrocorticosterone acetate," *J. Am. Chem. Soc.* 68:2478–2483 (1946).

10. P. K. Hench, "A Nobel prize story," *Patient Care* 33:173-74 (1999).

11. http://www.nap.edu/html/biomems/pjulian.html

12. http://www.blackinventor.com/pages/percyjulian.html

THE CASE OF THE FLOPPY-EARED RABBITS

1. B. Barber and R. Fox, "The case of the floppy-eared rabbits: An instance of serendipity gained and serendipity lost." *Am. J. Sociology* 64:128–36 (1958).

2. *Ibid.*, p. 132.

3. L. Thomas, "Reversible collapse of rabbit ears after intravenous papain, and prevention of recovery by cortisone," *J. Exper. Med.* 104:245–252 (1956).

4. Harold M. Schmeck, Jr., "Medical; Research Project Accidentally Wilts the Ears of a Rabbit," *New York Times,* February 20, 1957, p. 35.

5. Miscellany, "A lop-eared lapin" *LIFE* magazine, (March 4, 1957) p. 138.

6. L. Thomas, "Recent advances in research on rheumatic fever," *Minnesota Medicine* 35: 1105–1110 (1952).

7. Barber and Fox, p. 133.

8. L. Thomas and R. A. Good, "Bilateral cortical necrosis of the kidneys in cortisone-treated rabbits following injection of bacterial toxins," *Proc. Soc. Exper. Bio. & Med.* 76: 604–608 (1951).

9. Barber and Fox, p. 126.

10. A. Kellner and T. Robertson, "Selective necrosis of cardiac and skeletal muscle produced experimentally by means of proteolytic enzyme solutions given intravenously," *J. Exper. Med.* 94:387–404 (1954).

11. Barber and Fox, p. 135.

12. S. A. Morley, *A Talent to Amuse: A Biography of Noël Coward* (Boston, Little, Brown and Co., 1985).

13. L. Thomas, *The Youngest Science* (New York: Viking Press, 1983), p. 158.

Einstein and Jimmy Mac

1. J. Bernstein, *The Life It Brings: One Physicist's Beginnings* (Boston: Houghton Mifflin, 1987).

2. http://oaks.nvg.org/sa5ra17.html#einstein-anecdotes

3. P. Frank, *Einstein: His Life and Time* (New York: Knopf, 1947), p. 252.

4. http://kuhttp.cc.ukans.edu/carrie/docs/texts/einstein.txt

5. R. T. Sylves, *The Nuclear Oracles: A Political History of the General Advisory Committee of the Atomic Energy Commission*, (Ames, Iowa: Iowa State University Press, 1987).

6. J. Shreeve, *The Genome War* (New York: Knopf, 2004), p. 41–42.

7. http://www.norfolka2z.co.uk/tv/roughton.htm

8. J. D. Watson, *Genes, Girls, and Gamow: After the Double Helix* (New York: Knopf, 2002), p. 1.

9. http://www.sciencephotogallery.co.uk/articles/DNA_howPhotoArticle.php

10. J. D. Watson, *The Double Helix: A Personal Account of the Discovery of the Structure of DNA* (New York: Atheneum, 1968).

11. D. Berlinski, "Lucky Jim," *The Weekly Standard* 7:41–42 Washington (March 18, 2002).

12. A. Pais, *Subtle is the Lord: The Science and the Life of Albert Einstein* (New York and Oxford: Oxford University Press, 1982), p. vii.

13. http://www.randomhouse.com/boldtype/0202/watson/interview.html

14. Pais, p. 454.

15. F. J. Dyson, *Disturbing the Universe* (New York: Harper and Row, 1979).

16. Pais, p. 319.

17. *Ibid.*, p. i.

18. J. D. Watson and A. Berry, *DNA: The Secret of Life* (New York: Knopf, 2003), p. xii–xiii.

19. *Ibid.*, p. 161.

20. R. Finn, "McEnroe's Theatrics Get Mixed Review," *New York Times*, January 23, 1990.

21. M. Feeney, "Romancing the Molecule," in *Genes, Girls, and Gamow,* James D. Watson Looks at Love and DNA, *Boston Globe,* February 27, 2002, p. G1.

22. Watson and Berry, p. 291.

23. *Ibid.*, p.18.

24. *Ibid.*, p. 58.

25. http://www.laskerfoundation.org/awards/kwood/watson/publications.html

26. G. Weissmann, *The Year of the Genome* (New York: Times Books, 2002), p. 200.

27. Watson and Berry, p. 396.

28. J. Loeb, *The Mechanistic Conception of Life: Biological Essays* (1912) (Cambridge: Harvard University Press, 1964 reprint). p.3.

Baumol's Curse

1. Y. D. Coble "A Healthy America"[Letter]; *New York Times,* New York, New York, July 13, 2003; Late Edition (East Coast), p. 4.12.

2. http://www.demog.berkeley.edu/~andrew/1918/figure2.html

3. http://www.ahcpr.gov/news/ulp/costs/ulpcosts1.htm

4. Peter Landers, Leading the News: "Industry Urges Action on Health Costs," *Wall Street Journal,* New York, New York, June 11, 2002; Eastern Edition, pg. A3.

5. S. J. Bernstein, H. Rigter, B. Brorsson, L. H. Hilborne, L. L. Leape, A. P. Meijler, J. K. Scholma, A. S. Nord, "Waiting for coronary revascularization: a comparison between New York State, The Netherlands and Sweden," *Health Policy* 42(1):15–27 (October 1997).

6. R. J. Blendon, C. Schoen, C. DesRoches, *et al.*, "Inequities in health care: a five-country survey," *Health Aff.* (Millwood) 42(1):15–27 (May-June, 2002).

7. W. J. Baumol and W. G. Bowen, *Performing Arts: The Economic Dilemma* (New York: The Twentieth Century Fund, 1966).

8. W. J. Baumol, "Children of Performing Arts, the Economic Dilemma: The Climbing Costs of Health Care and Education," *Journal of Cultural Economics* (20):183–206 (1996).

9. W. J. Baumol, "Social Wants and Dismal Science: The Curious Case of the Climbing Costs of Health and Teaching," *Proceedings of the American Philosophical Society* 137(4)612–37 (1993).

10. G. Weissmann, *The Doctor Dilemma* (New York: Whittle Books, Grand Rounds Press, 1992), pp. 1–69.

11. H. Zinsser, *As I Remember Him: The Biography of R.S.* (Boston: Little Brown, 1940), p. 115.

12. S. Glied and J. G. Zivin, "How do doctors behave when some (but not all) of their patients are in managed care?" *J. Health Econ.* 21(2):337–53 (March 2002).

13. G. Flaubert, *Madame Bovary* (New York: Penguin Classics, 1984), p. 102.

From the Patchwork Mouse to Patchwork Data

1. J. Hixson, *The Patchwork Mouse* (Garden City, New York: Anchor Press/Doubleday, 1976), pp. 4–5.

2. J. Brody, "Charge of False Research Data Stirs Cancer Scientists at Sloan-Kettering," *New York Times,* April 18, 1974, p. 20.

3. W. T. Summerlin, "Allogeneic Transplantation of organ cultures of adult human skin," *Clin. Immunol. Imunopath.* 1:372–384 (1973).

4. W. T. Summerlin, G. E. Miller, R. A. Good, "Successful tissue and organ transplantation without immunosuppression," *J. Clin. Ivest.* 52:34a (1973).

5. W. S. Hwang, S. I. Roh, B. C. Lee, S. K. Kang, D. K. Kwon, *et al.* "Patient-specific embryonic stem cells derived from human SCNT blastocysts," *Science* 308:1777–1783 (2005).

6. Summary of the Final Report on Hwang's Research Allegation. Seoul National University Investigation Committee, http://www.snu.ac.kr/engsnu/- Accessed January 2006.

7. Gina Kolata, "Clone Scandal: 'A Tragic Turn' for Science," *New York Times,* December 16, 2005, p. A6.

8. Nicholas Wade, "It May Look Authentic; Here's How to Tell It Isn't," *New York Times,* January 24, 2006 (Science), p. 1.

9. D. Kennedy, "Good News—and Bad," (Editorial) *Science* 311:145 (2006).

10. J. Sudbø, J. J. Lee, S. M. Lippman, J. Mork, *et al.*, "Non-steroidal anti-inflammatory drugs and the risk of oral cancer: a nested case-control study," *The Lancet* 366:1359–68 (2005).

11. http://www.the-scientist.com/news/display/22952-

12. G. D. Curfman, S. Morrissey, J. M. Drazen, "Expression of Concern," *N. Engl. J. Med.* 350:1405–13 (2004), *N. Engl. J. Med.*, January 20, 2006 [E-pub ahead of print].

13. G. Cook, "Medical Journal Says Papers May be Fraudulent," *Boston Globe,* January 21, 2006, p. A9.

14. J. Couzin, "Scientific misconduct. MIT terminates researcher over data fabrication," *Science* 2005 310:758.

15. http://www.newscientist.com/article.ns?id=dn8230

16. L. Van Parijs, D. A. Peterson, A. K. Abbas, "The Fas/Fas ligand pathway and Bcl-2 regulate T cell responses to model self and foreign antigens," *Immunity* 8:265–74, (1998).

17. L. Van Parijs, Y. Refaeli, J. D. Lord, B. H., A. K. Abbas, D. Baltimore, "Uncoupling IL-2 signals that regulate T cell proliferation, survival, and Fas-mediated activation-induced cell death," *Immunity* 11:281–8 (1999).

18. E. Marshall, "Fraud Strikes Top Genome Lab," *Science* 274:908–910 (November 8, 1996).

19. http://ori.dhhs.gov/documents/newsletters/vol5_no4.pdf-

20. A. Hajra, P. P. Liu, N. A. Speck, F. S. Collins, "Overexpression of core-binding factor alpha (CBF alpha) reverses cellular transformation by the CBF beta-smooth muscle myosin heavy chain chimeric oncoprotein," *Mol. Cell Biol.* 15: 4980–9 (1995).

21. S. Mader, H. Lee, A. Pause, N. Sonenberg, "The translation initiation factor eIF-4E binds to a common motif shared by the translation factor eIF-4 gamma and the translational repressors 4E-binding proteins," *Mol. Cell Biol.* 15: 4990–7 (1995).

Alice James and Rheumatic Gout

1. A. R. Burr, ed., *Alice James: Her Brothers–Her Journal* (New York: Dodd, Mead, 1934), p. 232.

2. J. Strouse, *Alice James: A Biography* (Boston: Houghton Mifflin, 1980), p. 207.

3. *Ibid.*, p. 234.

4. A. B. Garrod, "Observations on certain pathological conditions of the blood and urine in gout, rheumatism and Bright's disease," *Med. Chir. Trans.* 31:83–97 (1848).

5. G. P. Rodnan, C. McEwen, S. L. Wallace [Cover story] "Sir Alfred Baring Garrod, FRS," *J. Am. Med. Assoc.* 224:663–5 (1973).

6. G. Weissmann and G. Rita, "The Molecular Basis of Gouty Inflammation: Interaction of Monosodium Urate Crystals with Lysosomes and Liposomes," *Nature New Biology* 240(101):167–172 (December 6, 1972).

7. A. B. Garrod, *A treatise on gout and rheumatic gout (rheumatoid arthritis),* 3rd ed., thoroughly rev. and enl. (London: Longmans, Green, 1876).

8. Strouse, p. 237.

9. R. Porter and G. S. Rousseau, *Gout: The Patrician Malady* (New Haven: Yale University Press, 1998).

10. R. Lagier, "Nosology versus pathology, two approaches to rheumatic diseases illustrated by Alfred Baring Garrod and Jean-Martin Charcot. *Rheumatology* 40:467–7 (2001).

11. A. B. Garrod, "Aix-les-Bains: the value of its course in rheumatoid arthritis, gout, rheumatism, and other diseases," *The Lancet* 869–71 (1889).

12. Strouse, p. 237.

13. S. Sontag, *Alice in Bed* (New York: Farrar, Straus and Giroux, 1993).

14. Burr, p. 161.

15. *Ibid.*, p. 252.

16. *Ibid.*, p. 230.

17. *Ibid.*, p. 87.

18. *Ibid.*, p. 227.
19. *Ibid.*, p. 230.
20. Strouse, p. 83.
21. Burr, p. 246
22. *Ibid.*, p. 242.
23. *Ibid.*, p. 247.
24. F. O. Matthiessen, *The James Family: Including Selections from the Writings of Henry James, Senior, William, Henry, and Alice James.* (New York: Knopf, 1961).
25. Strouse, p. 126.
26. Burr, p. 249.
27. F. Darwin, *The Life and Letters of Charles Darwin* (Cambridge: 1887; reprint, New York: Basic Books, 1959).
28. A. Desmond and J. Moore, *Darwin: The Life of a Tormented Evolutionist* (New York: Warner, 1991).
29. Strouse, p. 118.
30. E. Shorter, *From Paralysis to Fatigue: A History of Psychosomatic Illness in the Modern Era* (New York: Free Press, 1993).
31. Strouse, p. 185.
32. *Ibid.*, p. ix.
33. S. Wessely, "Medically unexplained symptoms: exacerbating factors in the doctor-patient encounter," *J R Soc Med* 96(5):223–7 (May 2003).
34. S. Wessely and M. Hotopf, "Is fibromyalgia a distinct clinical entity? Historical and epidemiological evidence," *Baillieres Best Pract. Res. Clin. Rheumatol* 13(3):427–36 (September 1999).
35. A. E. Garrod, "Inborn errors of metabolism," *The Lancet* 2:1–7, 73–79, 142–148, 214–220 (1908).
36. A. Pardee, F. Jacob, J. Monod, The genetic control and cytoplasmic expression of "inducibility" in the synthesis of ß-galactosidase by E. coli. *J. Mol. Biol.* 1:165 (1959).
37. E. Showalter, *Hysteries* (New York: Columbia University Press, 1998).
38. C. Schine, *Alice in Bed* (New York: Knopf, 1983).
39. Sontag, pp. 1, 115.

Lewis Thomas and the Two Cultures

1. G-L. L. Buffon, *Discourse sur le Style* [DISCOURS PRONONCE A L'ACADEMIE FRANCAISE PAR M. DE BUFFON LE JOUR DE SA RECEPTION LE 25 AOUT 1753] *Texte de l'édition de l'abbé J. Pierre* (Paris: Librairie Ch. Poussielgue,)1896, p 5.
2. C. P. Snow, *Two Cultures and the Scientific Revolution* (Cambridge: Cambridge University Press, 1959), p. 10.
3. L. Thomas, *The Youngest Science* (New York: Viking, 1983) pp. 14–22.
4. H. Zinsser, *As I Remember Him: The Biography of R.S.* (Boston: Little, Brown and Co., 1940), p. 102.
5. L. Thomas, "Ceti," in *The Lives of a Cell* (New York: Viking Press, 1974), p. 45.
6. L. Thomas, "The Art and Craft of Memoir," in *The Fragile Species* (New York: Scribner's, 1992), p. 16.
7. L. Thomas, "On Various Words," in *The Lives of a Cell* (New York: Viking Press, 1974), p. 128.
8. J. Lovelock, *GAIA: A New Look at Life on Earth* (Oxford: Oxford University Press, 1979).
9. J. Barzun, *A Stroll with William James* (Chicago: University of Chicago Press, 1983), p. 163.
10. L. Thomas, *Late Night Thoughts on Listening to Mahler's Ninth Symphony* (New York and London: Penguin Books, 1995), pp. 16–17.

11. E. Waugh, *A Little Order*, ed. Donat Gallagher, (Boston: Little, Brown & Co., 1977), p. 124.

12. L. Thomas, "Some Biomythology," in *The Lives of a Cell* (New York: Viking Press, 1974), p. 126.

13. L. Thomas, "Endotoxin," in *The Youngest Science* (New York: Viking Press, 1983), p. 151.

14. L. Thomas, "Limitation," *The Atlantic Monthly* 174:119 (1944).

15. E. Waugh, *A Little Order*, ed. Donat Gallagher (Boston: Little, Brown and Co., 1977), p. 110.

16. A. Silverstein, *A History of Immunology* (New York: Academic Press, 1989), p. 159.

17. S. J. Gould, "Calling Dr. Thomas," in *An Urchin in the Storm: Essays about Books and Ideas* (New York and London: W. W. Norton, 1987), p. 190.

18. S. Hook, "How to Blow Your Own Horn Effectively," *Wall Street Journal*, New York, November 23, 1987.

19. Christopher Lehmann-Haupt, [review] "The Fragile Species," *New York Times*, New York, April 16, 1992.

20. E. Waugh, *A Little Order*, ed. Donat Gallagher (Boston: Little, Brown and Co., 1977), p. 112.

Rats, Lice and Zinsser

1. H. Zinsser, *Rats, Lice and History* (Boston: Little, Brown and Co., 1935), pp. 13–14.

2. *Ibid.*, p. 185.

3. W. C. Summers, H. Zinsser, "A Tale of Two Cultures," *Yale J. Biol. & Med.* 72:341–347 (1999)

4. http://www.amazon.com/exec/obidos/tg/detail/-/0316988960/qid=1088431521/sr=8-1/ref=sr_8_xs_ap_i1_xgl14/104-5337232-0049523?v=glance&s=books&n=507846

5. H. Zinsser, *As I Remember Him: The Biography of R.S.*, (Boston: Little, Brown and Co., 1940), p. 102.

6. *Ibid.*, p. 116.

7. *Ibid.*, p. 129.

8. *Ibid.*, p. 65.

9. *Ibid.*, p. 73.

10. *Ibid.*, p. 142.

11. W. C. Wiliams, *The Autobiography of William Carlos Williams* (New York: Random House, 1951).

12. H. Zinsser, *As I Remember Him*, p. 146.

13. J. R. Lowell, "Commemoration Ode," in *Complete Poetical Works* (Boston: Houghton Mifflin, 1897).

14. H. Zinsser, *As I Remember Him*, p. 131.

15. *Ibid.*, p. 15.

16. S. B. Wolbach, Hans Zinsser, *Biogr. Mem. Nat'l Acad. Sci.* 24:323–360 (1948).

17. *Ibid.*, p. 8.

18. J. C. Aub and R. K. Hapgood, *Pioneer in Modern Medicine: David Linn Edsall of Harvard* (Boston: Harvard Medical Alumni Association, 1970), p. 324.

19. J. C. White, *quoted in* Tilton, E. M., *Amiable Autocrat: A Biography of Dr. Oliver Wendell Holmes* (New York: Henry Schuman, 1947), p. 223.

20. J. C. Aub, and R. K. Hapgood, p. 328.

21. B. Sicherman, *Alice Hamilton: A Life in Letters* (Cambridge: Harvard University Press, 1984), p. 237.

22. M. Muller, *Anne Frank: The Biography*, Robert Kimber (translator), Rita Kimber (translator) (New York: Henry Holt, 1998).

23. D. Raoult and V. Roux, "The body louse as a vector of re-emerging human diseases," *Clin. Infect. Dis.,* 888–911 (1999).

24. D. Raoult, V. Roux, J. B. Ndihokubwayo, G. Bise, D. Baudon, G. Marte, R. Birtles, "Jail fever (epidemic typhus) outbreak in Burundi," *Emerg. Infect. Dis.*, 357–60 (1997).

25. H. Zinsser and F. B. Grinell, *J. Immunol.* 10: 725–730 (1925).

26. A. H. Coons and M. Kaplan, "Localization of antigen in tissue cells. II. Improvements in a method for the detection of antigen by means of fluorescent antibody," *J. Exp. Med.* 91:1–13 (1950).

27. H. Zinsser, *As I Remember Him*, p. 293.

28. *Ibid.*, p. 441.

29. H. Zinsser, *Rats, Lice and History*, p. 14.

Reducing the Genome

1. H. Zinsser, *As I Remember Him: The Biography of R.S.* (Boston: Little Brown & Co., 1940), p. 116.

2. J. Watson and F. H. Crick, "Molecular structure of nucleic acids: a structure for deoxyribose nucleic acid," *Nature* 171(737):737–8 (April 26, 1953).

3. I. Wilmut, L. Young, K. H. Campbell, "Embryonic and somatic cell cloning," *Reprod. Fertil. Dev.* 10(7–8):639–43 (1998).

4. J. C. Venter, *et al.,* "The Sequence of the Human Genome," *Science* 291: 1304–1351 (2001).

5. E. Lander, *et al.,* "The Genome International Sequencing Consortium. Initial sequencing and analysis of the human genome," *Nature* 409:860–921 (2001).

6. J. W. Draper, *The Conflict Between Religion and Science* (New York: Appleton, 1874).

7. G. Liles "God's work in the lab." *MD Magazine* 32:49–53 (1992).

8. C. Venter "Sequencing the Human Genome." Lecture at the Marine Biological Laboratory, Woods Hole, Massachusetts, August 17, 2001: http://www.mblwhoilibrary.org /services/lecture_series/venter/

9. S. Brenner, "Hunting the Metaphor," *Science* 291:1265–1266 (2001).

10. J. Loeb, *The Mechanistic Conception of Life*, reprint of 1912 original (Cambridge: Harvard University Press, 1965).

11. G. Weissmann (editor) *The Biological Revolution: Contributions of Cell Biology to the Public Welfare* (New York: Plenum Press, 1979).

12. J. Shreeve, *The Genome War* (New York: Knopf, 2004).

13. R. Darnton, *The Business of Enlightenment: A Publishing History of the Encyclopedia, 1775–1800* (Cambridge: Harvard University Press, 1979).

14. P. N. Furbank, *Diderot: A Critical Biography* (New York: Knopf, 1992).

15. W. McKibben, "Unlikely Allies Against Cloning," *New York Times,* March 27, 2002, p. 23.

16. Editorial. *Harper's New Monthly Magazine.* (1853–1854) 8:690.

17. Draper, p. vi.

18. J. R. Wilmoth, "The Future of Human Longevity: A Demographer's Perspective," Science Online. In The Department of Demography at the University of California Berkeley. http://www.demog.berkeley.edu/~jrw/Papers/science.html

19. D. Fleming, *John William Draper* (Philadelphia: University of Pennsylvania Press, 1950), also "CENTENARY OF MAKER OF FIRST PORTRAIT PHOTOGRAPH"; "New York University Will Honor the Memory of Prof. John William Draper, Who Took the First Human Likeness When Daguerre Failed to Do It," *New York Times,* April 30, 1911, p. 14 Sunday Magazine.

20. J. W. Draper, *Human Physiology. Statical and Dynamical; or, the Conditions and Course*

of the Life of Man (New York: Harper & Brothers, 1856), p. 555.

21. J. W. Draper, *New-York Historical Society* [report] *New York Times,* December 2, 1864. p. 1.

22. K. S. Thomson. "Huxley, Wilberforce and the Oxford Museum," *American Scientist,* 88:210–215 (2002).

23. O. Chadwick, *The Secularization of the European Mind in the 19th Century* (Cambridge: University Press, 1975), p. 162.

24. R. Hooke, *Preface to Micrographia* (1665; Reprint, New York: Dover 1962), p. xi.

25. F. Bruni, "Pope Says Modern Mankind Is Usurping 'God's Place'" *New York Times,* August 19, 2002, p. 3.

26. J. G. Pickering, S. Takeshita, L. Feldman, D. W. Losordo, J. M. Isner, "Vascular applications of human gene therapy," [Review] *Semin. Interv. Cardiol.* 1(1):84–8 (March 1996).

27. D. Diderot, "*Rève d'Alembert*" in *Le Neveu De Rameau,* Cartonnage *éditeur LE CLUB FRANCAIS DU LIVRE, PARIS* (1947), p. 317.

28. Pierre-Jean-George Cabanis, *Rapports du Physique et du Moral de l'Homme* (Paris: Bechet Jeune, 1824).

29. G. Rizzolatti, L. Craighero, "The mirror-neuron system." *Annu. Rev. Neurosci.* 27:169–92 (2004).

30. E. R. Kandel, Nobel Lecture. "The Molecular Biology of Memory Storage: A Dialog between Genes and Synapses." http://nobelprize.org/medicine/laureates/2000/kandel-lecture.html

31. J. Moleschott, *Lehre des Nahrensmittel. Für das Volk* (Erlangen: University Ausgabe, 1850).

32. P. Greengard, Nobel Lecture, "The Neurobiology of Dopamine Signaling." http://nobelprize.org/medicine/laureates/2000/greengardlecture.html

33. J. Smuts, *Holism and Evolution* (London: Macmillan, 1926).

34. H. Taine, *de l'Intelligence,* (T. D. Haye translation of 1872 original. (New York: Henry Holt and Company, 1879), p. 34.

35. *Ibid.*, p. ix.

36. Jules-Antoine Castagnary, *Le Siecle* (April 29, 1874).

37. Taine, *op. cit.*, p. 36.

A Nobel Error

1. http://www.nobel.se/medicine/laureates/1954/

2. http://www.nobel.se/medicine/laureates/1934/

3. H. K. Beecher and M. D. Altschule, *Medicine at Harvard: The First 300 Years,* (Dartmouth, New Hampshire: New England Universities Press, 1977), p. 304.

4. *Ibid.*

5. G. S. Whipple and F. S. Robscheit Robbins, "Blood regeneration in severe anemia. Favorable influence of liver, heart and skeletal muscle in diet," *Am. J. Physiol.* 78:408–418 (1925).

6. G. R. Minot and W. P. Murphy, "Observations on patients with pernicious anemia partaking of a special diet. A clinical aspect," *Trans. Ass. Amer. Phys.* 41:72–5 (1926).

7. G. R. Minot, E. J. Cohn, W. P. Murphy, and H. A. Lawson, "Treatment of pernicious anemia with liver extract: effects upon the production of immature and mature red cells," *Am. J. Med. Sci.* 175:599–622 (1928).

8. F. W. Peabody, "The Pathology of the Bone Marrow in Pernicious Anemia," *Amer. J. Path.* 3:179–202 (1927).

9. *Ibid.*

10. W. P. Castle, "The Conquest of Pernicious Anemia," in *Blood Pure and Eloquent*, M. Wintrobe, ed. (New York: McGraw-Hill, 1980), p. 297.

11. L. S. Kass, *Pernicious Anemia* (Philadelphia: WB Saunders Company, 1976).

THE MOTHER OF US ALL: BOSTON CITY AND THE THORNDIKE

1. C. Caustic (nom de plume), *Terrible Tractoration: A Poetical Petition Against Galvanizing Trumpery. . . .* 1804 quoted in Holmes, O. W., "Homeopathy and Kindred Delusions," 1842 in *Medical Essays* (Boston: Riverside Press, 1892 edition), p. 35.

2. M. Finland and W. B. Castle, eds., *The Harvard Medical Unit at the Boston City Hospital* (Boston: Countway Library of Medicine HMS, 1983), p. 336. [hereafter: *HMU/BCH*]

3. H. G. Clark, *Outlines of a Plan for a Free City Hospital* (Boston: George C. Rand & Avery, 1860), p. 112, quoted in Charles Rosenberg, *The Care of Strangers: The Rise of America's Hospital System* (New York: Basic Books, 1987), p. 136.

4. Charles Rosenberg, *The Care of Strangers: The Rise of America's Hospital System* (New York: Basic Books, 1987), p. 339.

5. M. Finland, *Ibid.*, p. xxvi.

6. L. Thomas, *The Youngest Science: Notes of a Medicine Watcher* (New York: Viking, 1983), p. 38. [hereafter: TYS]

7. Rosenberg, p. 339.

8. W. L. Peltz, HMU/BCH, p. 297.

9. L. Thomas, TYS, p. 40.

10. F. J. Ingelfinger, HMU/BCH, p. 3I3.

11. L. Thomas, TYS, p. 36.

12. R. S. Evans, HMU/BCH, p. 353.

13. G. Weissmann, *Democracy and DNA* (New York: Knopf, 1995), p. 36.

14. L. Thomas, HMU/BCH, p. 337.

15. H. S. Lawrence, Presentation of the George M. Kober Medal to Lewis Thomas, Trans. Association of American Physicians 96: cxviii–cxxxiv (1983).

16. R. V. Ebert, HMU/BCH, p. 304.

17. W. L. Peltz, HMU/ BCH, p. 298.

18. F. W. Peabody, *Doctor and Patient* (New York: Macmillan, 1931).

19. G. Weissmann, "Against *Aequanimitas*," in *The Woods Hole Cantata* (New York: Dodd, Mead, 1985), p. 211.

20. F. W. Peabody, "The Care of the Patient," *Journal of the American Medical Association* 88:877–882 (1927).

21. J. C. Aub and R. K. Hapgood, *Pioneer in Modern Medicine: David Linn Edsall of Harvard* (Boston: Harvard Medical Alumni Association, 1970), p. 279.

22. H. K. Beecher and M. D. Altschule, *Medicine at Harvard: The First 300 Years* (Dartmouth, New Hampshire: New England Universities Press, 1977), pp. 303–307.

23. W. B. Bean, HMU/BCH, p. 267.

24. A. Langmuir, HMU/BCH, p. 252.

25. W. B. Bean, HMU/BCH, p. 268.

26. P. Kunkel, HMU/BCH, p. 323.

27. *Ibid.*, p. 322.

28. J. C. Owen, HMU/BCH, p. 140.

29. L. Thomas, TYS, p. 60.

Childish Curiosity

1. http://nihroadmap.nih.gov

2. http://grants1.nih.gov/grants/guide/notice-files/NOT-RM-05-011.html

3. L. Thomas, *The Lives of a Cell* (The Planning of Science) (New York: Viking Press, 1974), pp. 116–7.

4. W. T. Astbury, "Molecular biology or ultrastuctural biology?" [Editorial] *Nature* 190:1124 (1961).

5. D. J. McCarty, "Crystal-induced inflammation of the joints," *Annu. Rev. Med.* 21:357–66 (1970).

6. G. Weissmann and S. Weissmann, "X-ray diffraction studies of human aortic elastin residues," *J. Clin. Invest.* 39:1657–1666 (1960).

7. E. Chargaff, "Chemical specificity of nucleic acids and mechanism of their enzymatic degradation," *Experientia* 6:201–209 (1950).

8. R. Yeo, *Defining Science: William Whewell; Natural Knowledge and Public Debate in Early Victorian Britain* (New York: Cambridge University Press, 1993).

9. R. Kurzrock and C. C. Lieb, "Biochemical studies of human semen. II. The action of semen on the human uterus," *Proc. Soc. Exp. Biol. Med.* 26:268–72 (1930).

10. J. R. Vane, "Inhibition of prostaglandin synthesis as a mechanism of action for aspirin-like drugs," *Nature New Biology* 231:232–3 (1971).

11. D. Lednicer, "Tracing the origins of COX-2 inhibitors' structures," *Curr. Med. Chem.* 15:1457–61 (2002).

12. L. Thomas, *The Lives of a Cell* (The Planning of Science) (New York: Viking Press, 1974), p. 102.

Jacques Loeb and Stem Cells

1. Anon: "Loeb Tells of Artificial Life" *Chicago Daily Tribune,* December 28, 1900, p.12.

2. http://www.whitehouse.gov/news/releases/2006/01/20060131-10.html

3. L. R. Kass, "Babies by means of in vitro fertilization: unethical experiments on the unborn?" *N. Engl. J. Med.* 285(21):1174–9 (November 18, 1971).

4. Anon: "Dr. Loeb's Incredible Discovery" *New York Times,* March 2, 1905, p. 8.

5. W. J. Smith, "Kass, in the Firing Line: they hate Bush's bioethics man, too." *National Review Online.* http://www.nationalreview.com/comment/smith200312050930.asp

6. K. Phillips, *American Theocracy, The Peril and Politics of Radical Religion, Oil, and Borrowed Money in the 21st Century* (New York: Viking/Penguin, 2006).

7. http://www.nap.edu/catalog/11278.html

8. http://www.wordreference.com/fren/escroc

9. Nicholas Wade, "Science Academy Creating Panel to Monitor Stem-Cell Research" *New York Times,* February 16, 2006.

10. http://bioethicsprint.bioethics.gov/reports/white_paper/press_conference.html

11. http://www.asrm.org/whatsnew.html

12. J. Loeb, "On Artificial Parthenogenesis in Sea Urchins," *Science* 11:612–614 (1900).

13. "Reproduction of Humans," *Boston Evening Transcript,* October 2, 1900.

14. W. A. Evans, "Prof. Loeb has Fatherless Frog," *Chicago Daily Tribune,* September 25, 1912, p.2.

15. Letter: RELIGION AND BIOLOGY. How the Search for the Origin of Life Is Hampered by Preconceptions, *New York Times,* March 5, 1905, pg. 8.

16. J. Loeb, *Comparative Physiology of the Brain and Comparative Psychology* (New York: G. P. Putnam's Sons, 1900), p. 287.

17. G. Pincus, *The Eggs of Mammals* (New York: Macmillan, 1936).

18. Brave New World, *New York Times,* March 28, 1936, p. 14.

19. J. D. Ratcliff, "No Father to Guide Them" *Collier's,* March 20, 1937, p.137, quoted in Asbell, B., *The Pill* (New York: Random House), p. 120.

Galton's Prayer

1. http://galton.org-/

2. F. Galton, "Statistical inquiries into the efficacy of prayer," *Fortnightly Review* 12:125–35 (1872).

3. H. Benson, J. A. Dusek, J. B. Sherwood, *et al.*, "Study of the Therapeutic Effects of Intercessory Prayer (STEP) in cardiac bypass patients: a multicenter randomized trial of uncertainty and certainty of receiving intercessory prayer," *Am. Heart J.* 151:934–42 (2006).

4. Benedict Carey, "Healing power of prayer debunked," *New York Times,* March 31, 2006, p. A1,6.

5. Mark Melady, "'God factor' defended; Prayer study flawed," *Telegram & Gazette,* April 13, 2006, p. A1.

6. Albert Lindsey, "Area residents challenging prayer study," *Richmond Times-Dispatch,* Richmond, Virginia, April 9, 2006, p. B1.

7. Oliver Burkeman, "If you want to get better—don't say a little prayer," *The Guardian,* London (UK), April 1, 2006, p. A2,15.

8. M. W. Krucoff, S. W. Crater, K. L. Lee, "From efficacy to safety concerns: a STEP forward or a step back for clinical research and intercessory prayer? The Study of Therapeutic Effects of Intercessory Prayer (STEP)," *Am. Heart J.* 151:762–4 (2006).

9. J. A. Dusek, J. B. Sherwood, R. Friedman, *et al.*, "Study of the Therapeutic Effects of Intercessory Prayer (STEP): study design and research methods," *Am. Heart J.* 143:577–84 (2002).

10. M. Light, "Prayer studies a waste of money," *Buffalo News,* Buffalo, New York, April 23, 2006, p. H3.

11. http://www.csicop.org/si/2001-03/fringe-watcher.html

12. E. F. Targ, E. G. Levine, "The efficacy of a mind-body-spirit group for women with breast cancer: a randomized controlled trial," *Gen. Hosp. Psychiatry* 24:238–48 (2002).

13. H. H. Garrison and R. E. Palazzo, "What's Happening to New Investigators?" *FASEB. J,* 20:1288–89 (2006).

14. http://scientific.thomson.com/press/2005/8282889/

15. http://nccam.nih.gov/news/newsletter/2005_winter/prayer.htm

Acknowledgments

Many of these essays are extensions of material that has appeared in other venues.

Versions of the essays, *The Endarkenment; Intelligent Design: Galileo and the Lynxes; Swift-Boating Darwin: Alternative and Complementary Science; Red Wine, Ortolans and Chondroitin Sulfate; From the Patchwork Mouse to Patchwork Data; Childish Curiosity; Jacques Loeb and Stem Cells;* and *Galton's Prayer,* have appeared as editorials in *The FASEB Journal, The Official Journal of the American Societies of Experimental Biology.* I would like to thank Jennifer Pesanelli, Priscilla Markwood, Cody Mooneyhan and Susan Moore at the FASEB Publications Office in Bethesda MD for their tireless enthusiasm and for making the "FJ" an engaged, and engaging journal of life science.

The essays *Galileo's Gout, Cortisone and the Burning Cross,* and *Lewis Thomas and the Two Cultures* were published in *Pharos, The Magazine of Alpha Omega Alpha Honor Medical Society,* and I am grateful to Edward D. Harris, Jr., MD, its managing editor, for heading a journal that continues the tradition of medicine as a learned profession.

Versions of *Rats, Lice and Zinsser* and *Baumol's Curse* has appeared in *Emerging Infectious Diseases* and *Science and Spirit* respectively, while *Reducing the Genome* appeared in the (sadly) last issue of *Partisan Review.*

Thanks are obviously due to my editor, Erika Goldman for her noble efforts at bringing this volume to print, and to Jerome Lowenstein and Martin Blaser who presided over the birth of the press that has published it: *semper sunt in flores!*

Finally: neither this book, nor any of my other published work would have seen the light of comprehensive day, were it not for the unflagging editorial efforts and good cheer of Andrea Cody, the Administrative Coordinator of the Biotechnology Study Center at the NYU School of Medicine.

Index